Study Guide to Accompany
Structure & Function of the Body

TENTH EDITION

Prepared by
LINDA SWISHER, RN, EdD
Sarasota County Technical Institute
Sarasota, Florida

 Mosby

St. Louis Baltimore Boston Carlsbad Chicago Naples New York Philadelphia Portland
London Madrid Mexico City Singapore Sydney Tokyo Toronto Wiesbaden

Dedicated to Publishing Excellence

A Times Mirror
Company

Vice President & Publisher: James M. Smith
Editor: Ronald E. Worthington, PhD
Developmental Editor: Jean Sims Fornango
Project Manager: Gayle May Morris
Production Editor: Karen M. Rehwinkel
Manufacturing Supervisor: Karen Lewis

Printed in the United States of America.

Mosby-Year Book, Inc.
11830 Westline Industrial Drive
St Louis, Missouri   63146

International Standard Book Number: 0-8151-8716-5

Preface

To the Instructor

This study guide is designed to help your students master basic anatomy and physiology. It works in two ways.

First, the section of the preface, titled "To the Student," contains detailed instructions on:

How to achieve good grades in anatomy and physiology

How to read the textbook

How to use the exercises in this study guide

How to use visual memory as a learning tool

How to use mnemonic devices as learning aids

How to prepare for an examination

How to take an examination

How to find out why questions were missed on an examination

Second, the study guide itself contains features that facilitate learning. These features include the following:

1. LEARNING OBJECTIVES, designed to break down the process of learning into small units. The questions in this study guide have been developed to help the student master the learning objectives identified at the beginning of each chapter in the text. The guide is also sequenced to correspond to key areas of each chapter. A variety of questions has been prepared to cover the material effectively and expose the student to multiple learning approaches.

2. CROSSWORD PUZZLES and WORD FINDS, to encourage the use of new vocabulary words and emphasize the proper spelling of these terms.

3. OPTIONAL APPLICATION QUESTIONS, particularly targeted for the health occupations student, but appropriate for any student of anatomy and physiology because they are based entirely on information contained within the chapter.

4. DIAGRAMS, with key features marked by numbers for identification. Students can easily check their work by comparing the diagram in the workbook with the equivalent figure in the text.

5. PAGE NUMBER REFERENCES in the answer sections. Each answer is keyed to the appropriate text page. Additionally, questions are grouped into specific topics that correspond to the text. Each major topic of the study guide provides references to specific areas of the text so that students having difficulty with a particular grouping of questions have a specific reference area to assist them with remedial work. This is of great assistance to both instructor and student because remedial work is made easier and more effective when the area of weakness is identified accurately.

These features should make mastery of the material a rewarding experience for both instructor and student.

To the Student

How to Achieve Good Grades in Anatomy and Physiology

This study guide is designed to help you be successful in learning anatomy and physiology. Before you begin using the study guide, read the following suggestions. Students who understand effective study techniques and who have good study habits are successful students.

How to Read the Textbook

Keep up with the reading assignments. Read the textbook assignment before the instructor covers the material in lecture. If you have failed to read the assignment beforehand, you will not grasp what the instructor is talking about in lecture. When you read, do the following:

1. As you finish reading a sentence, ask yourself if you understand it. If you do not, put a question mark in the margin by that sentence. If the instructor does not clear up the problem in lecture, ask him or her to explain it to you.
2. Do the learning objectives in the text. A learning objective is a specific task that you are expected to be able to do after you have read a chapter. It sets specific goals for the student and breaks down learning into small steps. It emphasizes the key points that the author is making in the chapter.
3. Underline the text and make notes in the margin to highlight key ideas, to mark something you need to reinforce at a later time, or to indicate things that you do not understand.
4. If you come to a word you do not understand, look it up in a dictionary. Write the word on one side of an index card and its definition on the other side. Carry these cards with you, and when you have a spare minute, use them like flash cards to learn multiplication tables. If you do not know how to spell or pronounce a word, you will have a hard time remembering it.
5. Carefully study each diagram and illustration as you read. Many students ignore these aids. The author included them to help students understand the material.
6. Summarize what you read. After finishing a paragraph, try to restate the main ideas. Do this again when you finish the chapter. Identify and review in your mind the main concepts of the chapter. Check to see if you are correct. In short, be an active reader. Do not just stare at a page or read it superficially.

Finally, attack each unit of learning with a positive mental attitude. Motivation and perseverance are prime factors in achieving successful grades. The combination of your instructor, the text, the study guide, and your dedicated work will lead to success for you in anatomy and physiology.

How to Use the Exercises in This Study Guide

After you have read a chapter and learned all the new words, begin working with the study guide. Read the overview of the chapter, which summarizes the main points.

Familiarize yourself with the Topics for Review section, which emphasizes the learning objectives outlined in the text. Complete the questions and diagrams in the study guide. The questions are sequenced to follow the chapter outline and headings and are divided into small sections to facilitate learning. A variety of questions is offered throughout the study guide to help you cover the material effectively. The following examples are among the exercises that have been included to assist you.

MULTIPLE CHOICE QUESTIONS

Multiple choice questions will have only one correct answer out of several possibilities for you to select. There are two types of multiple choice questions that you may not be acquainted with:

1. "None of the above is correct" questions. These questions test your ability to recall rather than recognize the correct answer. You would select the "none of the above" choice only if all the other choices in that particular question were incorrect.
2. Sequence questions. These questions test your ability to arrange a list of structures in the

correct order. In this type of question you are asked to determine the sequence of the structures given in the various choices, and then you are to select the structure listed that would be third in that sequence, as in this example.

Which one of the following structures would be the third through which food would pass?
a. Stomach   d. Esophagus
b. Mouth   e. Anus
c. Large intestine

The correct answer would be a.

MATCHING QUESTIONS

Matching questions ask the student to select the correct answer from a list of terms and to write the answer in the space provided.

TRUE OR FALSE

True or false questions ask you to write T in the answer space if you agree with that statement. If you disagree with the statement, you will circle the incorrect word(s) and write the correct word(s) in the answer space.

IDENTIFY THE INCORRECT TERM

In questions that ask you to identify the incorrect term, three words are given that relate to each other in structure or function, and one more word is included that has no relationship, or has an opposing relationship to the other three terms. You are to circle the term that does not relate to the other terms. An example might be: iris, cornea, stapes, retina. You would circle the word stapes because all other terms refer to the eye.

FILL IN THE BLANKS

Fill-in-the-blank questions ask you to recall missing word(s) and insert it or them into the answer blank(s). These questions may be sentences or paragraphs.

APPLICATION

Application questions ask you to make judgments about a situation based on the information in the chapter. These questions may ask you how you would respond to a situation or to suggest a possible diagnosis for a set of symptoms.

CHARTS

Several charts have been included that correspond to figures in the text. Areas have been omitted so that you can fill them in and test your recall of these important areas.

IDENTIFICATION

The study guide includes word find puzzles that allow you to identify key terms in the chapter in an interesting and challenging way.

CROSSWORD PUZZLES

Vocabulary words from the New Words section at the end of each chapter of the text have been developed into crossword puzzles. This not only encourages recall, but also proper spelling. Occasionally, an exercise uses scrambled words to encourage recall and spelling.

LABELING EXERCISES

Labeling exercises present diagrams with parts that are not identified. For each of these diagrams you are to print the name of each numbered part on the appropriately numbered line. You may choose to further distinguish the structures by coloring them with a variety of colors. After you have written down the names of all the structures to be identified, check your answers. When it comes time to review before an examination you can place a sheet of paper over the answers that you have already written on the lines. This procedure will allow you to test yourself without peeking at the answers.

After completing the exercises in the study guide, check your answers. If they are not correct, refer to the page listed with the answer, and review it for further clarification. If you still do not understand the question or the answer, ask your instructor for further explanation.

If you have difficulty with several questions from one section, refer to the pages given at the end of the section. After reviewing the section, try to answer the questions again. If you are still having difficulty, talk to your instructor.

How to Use Visual Memory

Visual memory is another important tool in learning. If I asked you to picture an elephant in your mind, with all its external parts labeled, you could do that easily. Visual memory is a powerful key to learning. Whenever possible, try to build a memory picture. Remember, a picture is worth a thousand words.

Visual memory works especially well with the sequencing of items, such as circulatory pathways and the passage of air or food. Students who try to learn sequencing by memorizing a list of words do poorly on examinations. If they forget one word in the sequence, then they will forget all the words after the forgotten one as well. However, with a memory picture you can pick out the important features.

How to Use Mnemonic Devices

Mnemonic devices are little jingles that you memorize to help you remember things. If you make up your own, they will stick with you longer. Here are three examples of such devices:

"On Old Olympus' towering tops a Finn and German viewed some hops." This one is used to remember the cranial nerves. Each word begins with the same letter as does the name of one of the nerves.

"C. Hopkins CaFe where they serve Mg NaCl." This mnemonic device reminds you of the chemical symbols for the biologically important electrolytes.

"Roy G. Biv." This mnemonic device helps you remember the colors of the visible light spectrum.

How to Prepare For an Examination

Prepare far in advance for an examination. Actually, your preparation for an examination should begin on the first day of class. Keeping up with your assignments daily makes the final preparation for an examination much easier. You should begin your final preparation at least three nights before the test. Last-minute studying usually means poor results and limited retention.

1. Make sure that you understand and can answer all of the learning objectives for the chapter on which you are being tested.
2. Review the appropriate questions in this study guide. Reviewing is something that you should do after every class and at the end of every study session. It is important to keep going over the material until you have a thorough understanding of the chapter and rapid recall of its contents. If review becomes a daily habit, studying for the actual examination will not be difficult. Go through each question and write down an answer. Do the same with the labeling of each structure on the appropriate diagrams. If you have already done this as part of your daily review, cover the answers with a piece of paper and quiz yourself again.
3. Check the answers that you have written down against the correct answers in the back of the study guide. Go back and study the areas in the text that refer to questions that you missed and then try to answer those questions again. If you still cannot answer a question or label a structure correctly, ask your instructor for help.
4. Ask yourself as you read a chapter what questions you would ask if you were writing a test on that unit. You will most likely ask yourself many of the questions that will show up on your examinations.
5. Get a good night's sleep before the test. Staying up late and upsetting your biorhythms will only make you less efficient during the test.

How to Take an Examination

THE DAY OF THE TEST

1. Get up early enough to avoid rushing. Eat appropriately. Your body needs fuel, but a heavy meal just before a test is not a good idea.
2. Keep calm. Briefly look over your notes. If you have prepared for the test properly, there will be no need for last-minute cramming.
3. Make certain that you have everything you need for the test: pens, pencils, test sheets, and so forth.
4. Allow enough time to get to the examination site. Missing your bus, getting stuck in traffic, or being unable to find a parking space will not put you in a good frame of mind to do well on the examination.

DURING THE EXAMINATION

1. Pay careful attention to the instructions for the test.
2. Note any corrections.
3. Budget your time so that you will be able to finish the test.
4. Ask the instructor for clarification if you do not understand a question or an instruction.
5. Concentrate on your own test paper and do not allow yourself to be distracted by others in the room.

HINTS FOR TAKING A MULTIPLE CHOICE TEST

1. Read each question carefully. Pay attention to each word.

2. Cross out obviously wrong choices and think about those that are left.
3. Go through the test once, quickly answering the questions you are sure of. Then go back over the test and answer the rest of the questions.
4. Fill in the answer spaces completely and make your marks heavy. Erase completely if you make a mistake.
5. If you must guess, stick with your first hunch. Most often, students will change right answers to wrong ones.
6. If you will not be penalized for guessing, do not leave any blanks.

HINTS FOR TAKING AN ESSAY TEST

1. Budget time for each question.
2. Write legibly and try to spell words correctly.
3. Be concise, complete, and specific. Do not be repetitious or long-winded.
4. Organize your answer in an outline, which helps not only the student but also the person who grades the test.
5. Answer each question as thoroughly as you can, but leave some room for possible additions.

HINTS FOR TAKING A LABORATORY PRACTICAL EXAMINATION

Students have a hard time with this kind of test. Visual memory is very important here. To put it simply, you must be able to identify every structure you have studied. If you are unable to identify a structure, then you will be unable to answer any questions about that structure.

Possible questions that could appear on an examination of this type might include:

1. Identification of a structure, organ, feature.
2. Function of a structure, organ, feature.
3. Sequence questions for air flow, passage of food, urine, and so forth.
4. Disease questions (for example, if an organ fails, what disease will result?).

How to Find Out Why Questions Were Missed on an Examination

AFTER THE EXAMINATION

Go over your test after it has been scored to see what you missed and why you missed it. You can pick up important clues that will help you on future examinations. Ask yourself these questions:

1. Did I miss questions because I did not read them carefully?
2. Did I miss questions because I had gaps in my knowledge?
3. Did I miss questions because I could not determine scientific words?
4. Did I miss questions because I did not have good visual memory of things?

Be sure to go back and learn the things you did not know. Chances are these topics will come up on the final examination.

Your grades in other classes will improve as well when you apply these study methods. Learning should be fun. With these helpful hints and this study guide you should be able to achieve the grades you desire. Good luck!

Acknowledgments

I wish to express my appreciation to the staff of Mosby, especially Ron Worthington and Jean Fornango, for opening this door. My continued admiration and thanks to Gary Thibodeau and Kevin Patton for another outstanding edition of their text. Their time and dedication to science education will hopefully create a better quality of health care for the future.

Special thanks to Randy Fagan for her perseverance and enthusiasm while transposing the written word to typed script.

A. Christine Payne and her computer combined efforts to produce crossword puzzles and word finds for the units. Her creativity added the variety necessary to stimulate the learning process.

To my mother and sisters Penny and Kim, my thanks for your assistance and support throughout this project and my life.

Finally, to Bill and to my daughter Amanda, this book is dedicated. You and I know the many reasons.

LINDA SWISHER, RN, EdD

Contents

CHAPTER **1**

An Introduction to the Structure and Function of the Body

A command of terminology is necessary for a student to be successful in any area of science. This chapter defines the terminology and concepts that are basic to the field of anatomy and physiology. Building a firm foundation in these language skills will assist you with all future chapters.

The study of anatomy and physiology involves the structure and function of an organism and the relationship of its parts. It begins with a basic organization of the body into different structural levels. Beginning with the smallest level (the cell) and progressing to the largest, most complex level (the system), this chapter familiarizes you with the terminology and the levels of organization needed to facilitate the study of the body as a part or as a whole.

It is also important to be able to identify and to describe specific body areas or regions as we progress in this field. The anatomical position is used as a reference position when dissecting the body into planes, regions, or cavities. The terminology defined in this chapter allows you to describe the areas efficiently and accurately.

Finally, the process of homeostasis is reviewed. This state of relative constancy in the chemical composition of body fluids is necessary for good health. In fact, the very survival of the body depends on the successful maintenance of homeostasis.

TOPICS FOR REVIEW

Before progressing to Chapter 2, you should have an understanding of the structural levels of organization; the planes, regions, and cavities of the body; the terminology used to describe these areas, and the concept of homeostasis as it relates to the survival of the species.

STRUCTURAL LEVELS OF ORGANIZATION

Match the term on the left with the proper selection on the right.

___D___ 1. Organism

___E___ 2. Cells

___A___ 3. Tissue

___C___ 4. Organ

___B___ 5. Systems

A. Many similar cells that act together to perform a common function

B. The most complex units that make up the body

C. A group of several different kinds of tissues arranged to perform a special function

D. Denotes a living thing

E. The smallest ìlivingî units of structure and function in the body

▷ *If you have had difficulty with this section, review pages 1 and 2.*

ANATOMICAL POSITION

Match the term on the left with the proper selection on the right.

___C___ 6. Body

___A___ 7. Arms

___E___ 8. Feet

___D___ 9. Prone

___B___ 10. Supine

A. At the sides

B. Face upward

C. Erect

D. Face downward

E. Forward

▷ *If you have had difficulty with this section, review pages 3-5.*

ANATOMICAL DIRECTIONS
PLANES OR BODY SECTIONS

Fill in the crossword puzzle on the next page.

11. Upper or above *superior*

12. Lower or below *inferior*

13. Horizontal plane *transverse*

14. Front (abdominal side) *ventral*

15. Toward the side of the body *lateral*

16. Toward the midline of the body *medial*

17. Farthest from the point of origin of a body point *distal*

Fill in the crossword puzzle.

11. Upper or above
12. Lower or below
13. Horizontal plane
14. Front (abdominal side)
15. Toward the side of the body
16. Toward the midline of the body
17. Farthest from the point of
    origin of a body point

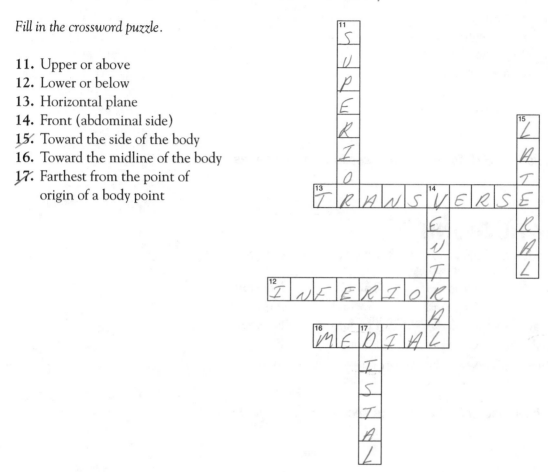

Circle the correct answer.
18. The stomach is (superior or (inferior)) to the diaphragm.
19. The nose is located on the ((anterior) or posterior) surface of the body.
20. The lungs lie (medial or (lateral)) to the heart.
21. The elbow lies ((proximal) or distal) to the forearm.
22. The skin is ((superficial) or deep) to the muscles below it.
23. A midsagittal plane divides the body into ((equal) or unequal) parts.
24. A frontal plane divides the body into ((anterior and posterior) or superior and inferior) sections.
25. A transverse plane divides the body into (right and left or (upper and lower)) sections.
26. A coronal plane may also be referred to as a (sagittal or (frontal)) plane.

▶ *If you have had difficulty with this section, review pages 3-7.*

BODY CAVITIES

Select the correct term from the choices given and insert the letter in the answer blank.

(A) ventral cavity   (B) dorsal cavity

_A_ 27. thoracic
_B_ 28. cranial
_A_ 29. abdominal
_A_ 30. pelvic
_A_ 31. mediastinum
_B_ 32. spinal
_A_ 33. pleural

▶ *If you have had difficulty with this section, review pages 5-6.*

BODY REGIONS

Circle the one that does not belong.

34. Axial	Head	Trunk	(Extremities)
35. Axillary	(Cephalic)	Brachial	Antecubital
36. Frontal	Orbital	(Plantar)	Nasal
37. (Carpal)	Crural	Plantar	Pedal
38. Cranial	Occipital	(Tarsal)	Temporal

▶ *If you have had difficulty with this section, review pages 8-10.*

THE BALANCE OF BODY FUNCTIONS

Fill in the blanks.

39. _Survival_ depends on the body's ability to maintain or restore homeostasis.
40. "Homeostasis" is the term used to describe the relative constancy of the body's _internal environment_
41. The basic type of homeostatic control system in the body is called a _feedback loop_ _____.
42. Homeostatic control mechanisms are categorized as either _negative_ or _positive_ feedback loops.
43. Negative feedback loops tend to _maintain, stabilize_ conditions.
44. Positive feedback control loops are _amplifiers stimulatory_
45. Changes and functions that occur during the early years are called _Developmental process_.
46. Changes and functions that occur after young adulthood are called _Aging process_.

▶ *If you have had difficulty with this section, review pages 8-12.*

APPLYING WHAT YOU KNOW

47. Mrs. Fagan has had an appendectomy. The nurse is preparing to change the dressing. She knows that the appendix is located in the right iliac inguinal region, the distal portion extending at an angle into the hypogastric region. Place an X on the diagram where the nurse will place the dressing.

48. Mrs. Suchman noticed a lump in her breast. Dr. Reeder noted on her chart that a small mass was located in the left breast medial to the nipple. Place an X where Mrs. Suchman's lump would be located.

49. Heather was injured in a bicycle accident. X-ray films revealed that she had a fracture of the right patella. A cast was applied beginning at the distal femoral region and extending to the pedal region. Place an X where Heather's cast begins and ends.

50. WORD FIND

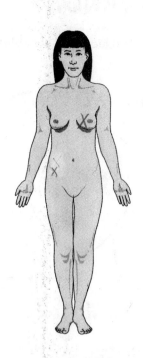

Can you find 18 terms from this chapter in the box of letters? Words may be spelled top to bottom, bottom to top, right to left, left to right or diagonally.

```
N  H  L  T  E  B  W  N  G  N  M  M  Y  X  A
O  O  H  A  V  U  U  C  L  W  E  P  N  G  L
I  M  U  N  I  T  S  A  I  D  E  M  W  A  L
T  E  R  T  V  C  T  S  I  C  J  S  R  T  K
A  O  N  O  P  T  I  A  I  W  A  T  Y  R  H
Z  S  Q  S  I  R  L  F  A  T  N  R  G  O  W
I  T  W  G  P  R  O  I  R  E  T  S  O  P  F
N  A  A  Z  J  E  E  X  V  E  E  H  L  H  Z
A  S  N  L  M  C  T  P  I  C  P  D  O  Y  T
G  I  C  A  Y  U  O  X  U  M  N  U  I  V  P
R  S  M  R  T  N  U  K  B  S  A  Y  S  V  M
O  R  C  U  R  O  I  R  B  S  Q  L  Y  U  G
L  H  X  E  M  P  M  E  T  S  Y  S  H  V  S
U  W  L  L  Q  D  U  Y  Y  N  E  E  P  J  B
Q  N  Z  P  K  D  B  O  D  C  G  I  N  J  A
```

DID YOU KNOW?

Many animals produce tears but only humans weep as a result of emotional stress.

DORSAL AND VENTRAL BODY CAVITIES

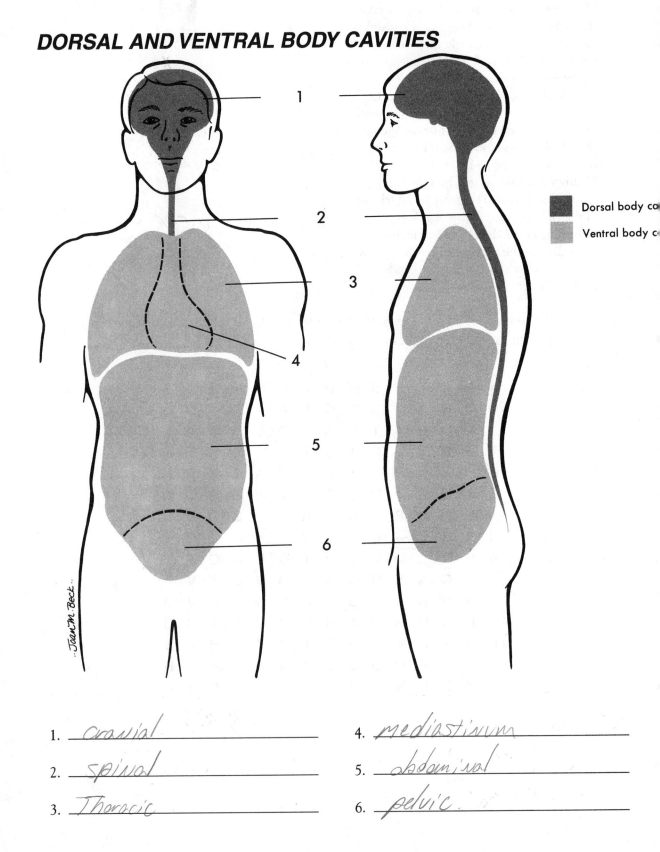

Dorsal body ca

Ventral body c

..JoanM.Beck..

1. _cranial_
2. _spinal_
3. _Thoracic_
4. _mediastinum_
5. _abdominal_
6. _pelvic_

DIRECTIONS AND PLANES OF THE BODY

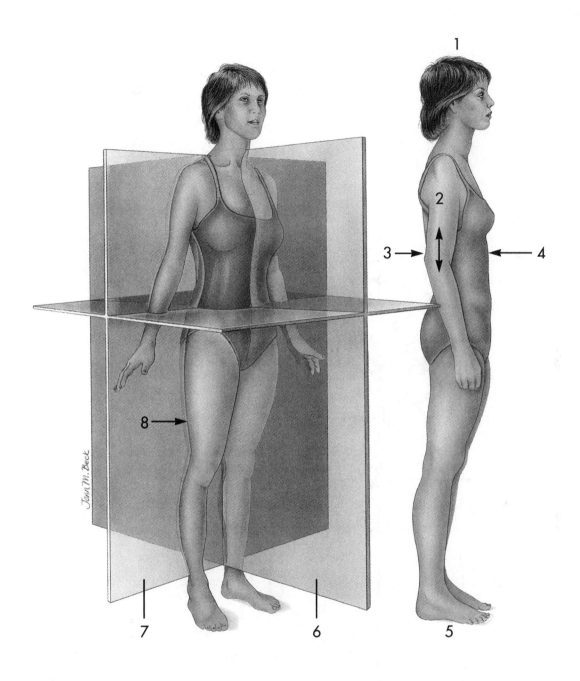

1. _Superior_
2. _Proximal_
3. _posterior (dorsal)_
4. _anterior (ventral_
5. _inferior_
6. _sagital plane_
7. _frontal plane_
8. _lateral_

REGIONS OF THE ABDOMEN

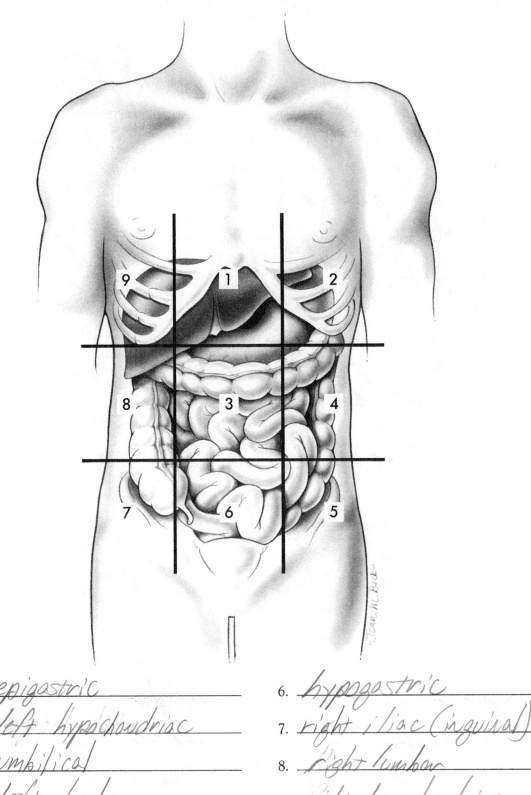

1. _epigastric_
2. _left hypochondriac_
3. _umbilical_
4. _left lumbar_
5. _left iliac (inguinal)_
6. _hypogastric_
7. _right iliac (inguinal)_
8. _Right lumbar_
9. _right hypochondriac_

CHAPTER **2**

Cells and Tissues

Cells are the smallest structural units of living things. Therefore, because we are living, we are made up of a mass of cells. Human cells can be viewed only microscopically. They vary in shape and size. The three main parts of a cell are the cytoplasmic membrane, the cytoplasm, and the nucleus. As you review the chapter on cells, you will be amazed at the cells' resemblance to the body as a whole. You will identify miniature circulatory systems, reproductive systems, digestive systems, power plants much like the muscular system, and many others that will aid in your understanding of these systems in future chapters.

Cells, just like humans, depend on water, food, gases, elimination of wastes, and numerous other substances to survive. The movement of these substances in and out of the cells is accomplished by two primary methods: passive transport processes and active transport processes. In passive transport processes no cellular energy is required to effect movement through the cell membrane. However, in active transport cellular energy is required to provide movement through the cell membrane.

Cell reproduction completes the study of cells. A basic explanation of DNA, "the hereditary molecule," gives us a proper respect for the capability of the cell to transmit physical and mental traits from generation to generation. Reproduction of the cell, called *mitosis*, is a complex process requiring several stages. These stages are outlined and diagrammed in the text to facilitate learning.

This chapter concludes with the discussion of tissues. The four main types of tissues (epithelial, connective, muscle, and nervous) are reviewed. Knowledge of the characteristics, location, and function of these tissues are necessary to complete your understanding of the next structural level of organization.

TOPICS FOR REVIEW

Before progressing to Chapter 3, you should have an understanding of the structure and function of the smallest living unit in the body—the cell. Your review should also include the methods by which substances are moved through the cell membrane and the stages necessary for the cell to reproduce. The study of this chapter is completed with an understanding of tonicity and body tissues and the function they perform in the body.

CELLS

Match the term on the left with the proper selection on the right.

Group A

_____	1. Cytoplasm
_____	2. Plasma membrane
_____	3. Cholesterol
_____	4. Nucleus
_____	5. Centriole

A. Component of plasma membrane
B. Controls reproduction of the cell
C. "Living matter"
D. Paired organelles
E. Surrounds cells

Group B

_____	6. Ribosomes
_____	7. Endoplasmic reticulum
_____	8. Mitochondria
_____	9. Lysosomes
_____	10. Golgi apparatus

A. "Power plants"
B. "Digestive bags"
C. "Carbohydrate producing and packaging factory"
D. "Protein factories"
E. "Smooth and rough"

Fill in the blanks.

11. The numerous small structures that function like organs in a cell are called

_____.

12. A procedure performed before transplanting an organ from one individual to another is

_____.

13. Fine, hairlike extensions found on the exposed or free surfaces of some cells are called

_____.

14. The process that uses oxygen to break down glucose and other nutrients to release energy required for cellular work is called _____ or _____ _____.

15. _____ are usually attached to rough endoplasmic reticulum and produce enzymes and other protein compounds.

16. The _____ provide energy-releasing chemical reactions that go on continuously.

17. The organelles that can digest and destroy microbes that invade the cell are called

_____.

18. Mucus is an example of a product manufactured by the _____.

19. These rod-shaped structures, _____, play an important role during cell division.

20. _____ are threadlike structures made up of proteins and DNA.

▷ *If you have had difficulty with this section, review pages 17-22.*

MOVEMENT OF SUBSTANCES THROUGH CELL MEMBRANES

Circle the correct choice.

21. The energy required for active transport processes is obtained from:
    A. ATP
    B. DNA
    C. Diffusion
    D. Osmosis

22. An example of a passive transport process is:
    A. Permease system
    B. Phagocytosis
    C. Pinocytosis
    D. Diffusion
23. Movement of substances from a region of high concentration to a region of low concentration is known as:
    A. Active transport
    B. Passive transport
    C. Cellular energy
    D. Concentration gradient
24. Osmosis is the _____ of water across a selectively permeable membrane.
    A. Filtration
    B. Equilibrium
    C. Active transport
    D. Diffusion
25. _____ involves the movement of solutes across a selectively permeable membrane by the process of diffusion.
    A. Osmosis
    B. Filtration
    C. Dialysis
    D. Phagocytosis
26. A specialized example of diffusion is:
    A. Osmosis
    B. Permease system
    C. Filtration
    D. All of the above
27. This movement always occurs down a hydrostatic pressure gradient.
    A. Osmosis
    B. Filtration
    C. Dialysis
    D. Facilitated diffusion
28. The uphill movement of a substance through a living cell membrane is:
    A. Osmosis
    B. Diffusion
    C. Active transport process
    D. Passive transport process
29. An example of an active transport process is:
    A. Ion pump
    B. Phagocytosis
    C. Pinocytosis
    D. All of the above
30. An example of a cell that uses phagocytosis is the:
    A. White blood cell
    B. Red blood cell
    C. Muscle cell
    D. Bone cell
31. A saline solution that contains a higher concentration of salt than living red blood cells would be:
    A. Hypotonic
    B. Hypertonic
    C. Isotonic
    D. Homeostatic
32. A red blood cell becomes engorged with water and will eventually lyse, releasing hemoglobin into the solution. This solution is _____ to the red blood cell.
    A. Hypotonic
    B. Hypertonic
    C. Isotonic
    D. Homeostatic

▷ *If you have had difficulty with this section, review pages 22-26.*

CELL REPRODUCTION

Circle the one that does not belong.

33. DNA	Adenine	Uracil	Thymine
34. Complementary base pairing	Guanine	RNA	Cytosine
35. Anaphase	Specific sequence	Gene	Base pairs

36. RNA	Ribosome	Thymine	Uracil
37. Translation	Protein synthesis	mRNA	Interphase
38. Cleavage furrow	Anaphase	Prophase	2 daughter cells
39. "Resting"	Prophase	Interphase	DNA replication
40. Identical	2 nuclei	Telophase	Metaphase
41. Metaphase	Prophase	Telophase	Gene

▶ *If you have had difficulty with this section, review pages 26-31.*

TISSUES

42. Fill in the missing area.

TISSUE	LOCATION	FUNCTION
Epithelial		
1. Simple squamous	1a. Alveoli of lungs	1a. _____
	1b. Lining of blood and lymphatic vessels	1b. _____
2. Stratified squamous	2a. _____	2a. Protection
	2b. _____	2b. Protection
3. Simple columnar	3. _____	3. Protection, secretion, absorption
4. _____	4. Urinary bladder	4. Protection
5. Pseudostratified	5. _____	5. Protection
6. Simple cuboidal	6. Glands; kidney tubules	6. _____
Connective		
1. Areolar	1. _____	1. Connection
2. _____	2. Under skin	2. Protection; insulation
3. Dense fibrous	3. Tendons; ligaments; fascia, scar tissue	3. _____
4. Bone	4. _____	4. Support, protection
5. Cartilage	5. _____	5. Firm but flexible support
6. Blood	6. Blood vessels	6. _____
7. _____	7. Red bone marrow	7. Blood cell formation
Muscle		
1. Skeletal (striated voluntary)	1. _____	1. Movement of bones
2. _____	2. Wall of heart	2. Contraction of heart
3. Smooth	3. _____	3. Movement of substances along ducts; change in diameter of pupils and shape of lens; "gooseflesh"
Nervous		
1. _____	1. _____	1. Irritability, conduction

▶ *If you have had difficulty with this section, review Table 2-5 and pages 32-40.*

APPLYING WHAT YOU KNOW

43. Mr. Fee's boat had capsized, and he was stranded on a deserted shoreline for 2 days without food or water. When found, he had swallowed a great deal of seawater. He was taken to the emergency room in a state of dehydration. In the space next to the word search below, draw the appearance of the red blood cells as they would appear to the laboratory technician.

44. The nurse was instructed to dissolve a pill in a small amount of liquid medication. As she dropped the capsule into the liquid, she was interrupted by the telephone. On her return to the medication cart, she found the medication completely dissolved and apparently scattered evenly throughout the liquid. This phenomenon did not surprise her since she was aware from her knowledge of cell transport that _____ had created this distribution.

45. Ms. Bence has emphysema and has been admitted to the hospital unit with oxygen per nasal cannula. Emphysema destroys the tiny air sacs in the lungs. These tiny air sacs, called *alveoli*, provide what function for Ms. Bence?

46. Merrily was 5'4" and weighed 115 lbs. She appeared very healthy and fit, yet her doctor advised her that she was "overfat." What might be the explanation for this assessment?

47. WORD FIND

Can you find 16 terms from this chapter? Words may be spelled top to bottom, bottom to top, right to left, left to right, or diagonally.

```
A  P  D  I  E  V  E  W  N  P  F  E  H  Y  X
I  G  I  Z  N  S  R  L  X  S  T  W  F  Z  S
R  J  V  N  N  T  A  I  L  M  R  P  H  M  F
D  W  U  E  O  O  E  H  B  E  A  D  U  G  F
N  W  I  U  I  C  I  R  P  O  N  N  F  B  W
O  P  U  R  S  H  Y  T  P  O  S  A  W  X  K
H  X  F  O  U  R  Y  T  A  H  L  O  G  A  W
C  L  Z  N  F  O  Y  P  O  R  A  E  M  R  M
O  U  I  X  F  M  L  F  O  S  T  S  T  E  O
T  T  B  Y  I  A  B  C  O  T  I  L  E  S  D
I  V  S  O  D  T  T  M  I  T  O  S  I  S  U
M  N  A  A  I  I  L  E  N  L  N  N  N  F  R
E  O  K  K  I  D  C  Q  H  B  I  V  I  T  H
E  K  T  X  B  Z  A  E  A  L  S  A  E  C  K
S  B  Z  R  P  M  V  L  X  W  Z  A  Z  C  V
```

Chromatid   Hypotonic   Pinocytosis
Cilia   Interphase   Ribosome
Cuboidal   Mitochondria   Telophase
DNA   Mitosis   Translation
Diffusion   Neuron
Filtration   Organelle

DID YOU KNOW?

The largest single cell in the human body is the female sex cell, the ovum. The smallest single cell in the human body is the male sex cell, the sperm.

CELLS AND TISSUES

ACROSS

2. Last stage of mitosis
5. Shriveling of cell due to water withdrawal
7. Fat
8. Cartilage cell
10. Ribonucleic acid (abbreviation)
12. Specialized example of diffusion
13. First stage of mitosis

DOWN

1. Having an osmotic pressure greater than that of the solution of which it is compared
3. Cell organ
4. Energy source for active transport
6. Nerve cell
9. Occurs when substances scatter themselves evenly throughout an available space
11. Reproduction process of most cells
12. Chemical "blueprint" of the body (abbreviation)

CELL STRUCTURE

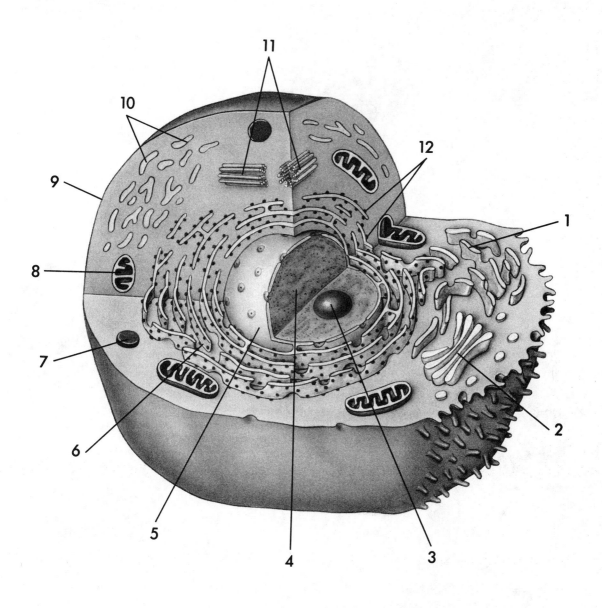

1. _____

2. _____

3. _____

4. _____

5. _____

6. _____

7. _____

8. _____

9. _____

10. _____

11. _____

12. _____

MITOSIS

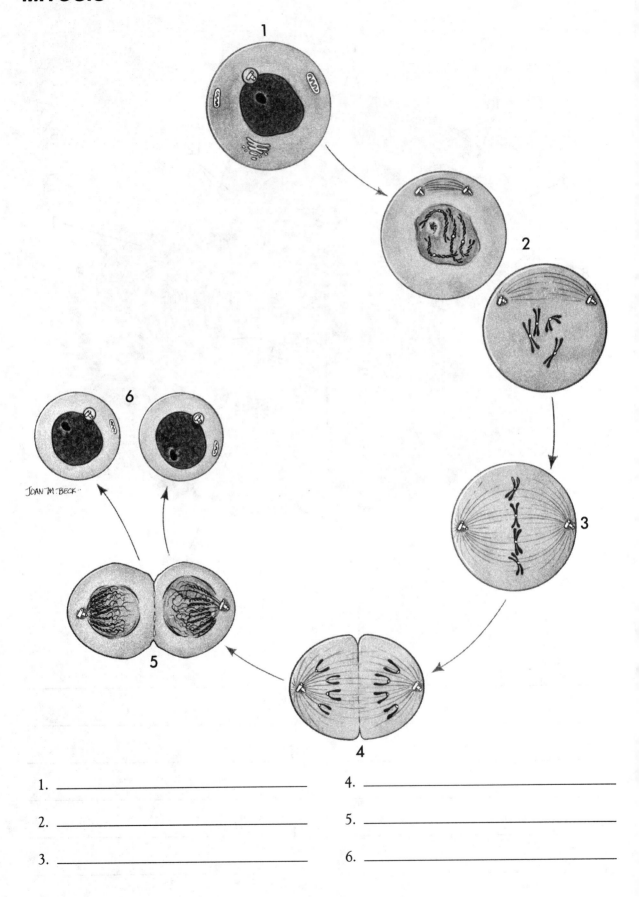

1. _____

2. _____

3. _____

4. _____

5. _____

6. _____

TISSUES

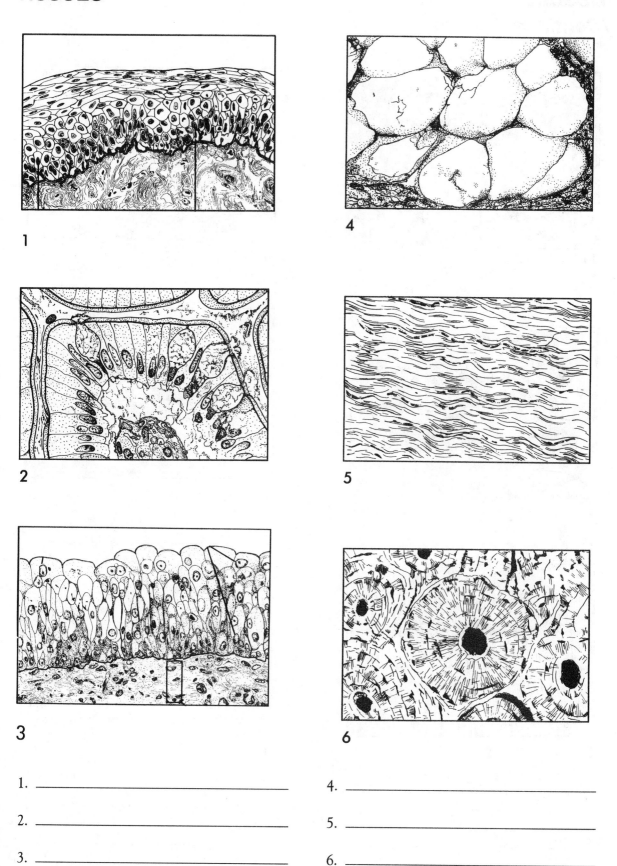

1

2

3

4

5

6

1. _____

2. _____

3. _____

4. _____

5. _____

6. _____

TISSUES
(Continued)

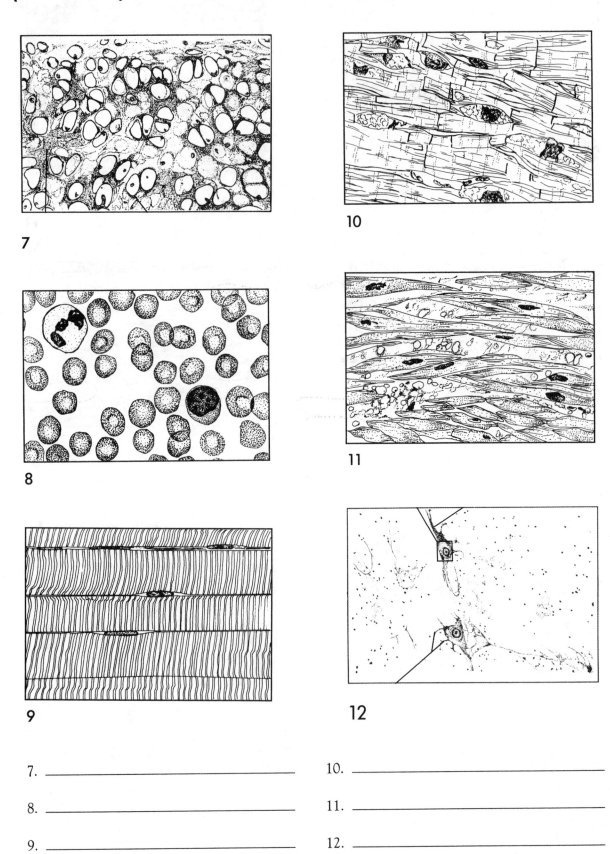

7

8

9

10

11

12

7. _____

8. _____

9. _____

10. _____

11. _____

12. _____

CHAPTER **3** # Organ Systems of the Body

A smooth-running automobile is the result of many systems working together harmoniously. The engine, the fuel system, the exhaust system, the brake system, and the cooling system are but a few of the many complex structural units that the automobile as a whole relies on to keep it functioning smoothly. So it is with the human body. We, too, depend on the successful performance of many individual systems working together to create and maintain a healthy human being.

When you have completed your review of the 11 major organ systems and the organs that make up these systems, you will find your understanding of the performance of the body as a whole much more meaningful.

TOPICS FOR REVIEW

Before progressing to Chapter 4, you should have an understanding of the 11 major organ systems and be able to identify the organs that are included in each system.

ORGAN SYSTEMS OF THE BODY

Match the term on the left with the proper selection on the right.

Group A

_A_	1. Integumentary	A. Hair
_E_	2. Skeletal	B. Spinal cord
_D_	3. Muscular	C. Hormones
_B_	4. Nervous	D. Tendons
_C_	5. Endocrine	E. Joints

Group B

_F_	6. Circulatory	— A. Esophagus
_E_	7. Lymphatic `	— B. Ureters
_B_	8. Urinary	— C. Larynx
_A_	9. Digestive	— D. Genitalia
_C_	10. Respiratory	E. Spleen
_D_	11. Reproductive	F. Capillaries

Circle the one that does not belong.

12. Pharynx	Trachea	(Mouth)	Alveoli
13. Uterus	(Rectum)	Gonads	Prostate
14. Veins	Arteries	Heart	(Pancreas)
15. (Pineal)	Bladder	Ureters	Urethra
16. Tendon	Smooth	(Joints)	Voluntary
17. (Pituitary)	Brain	Spinal cord	Nerves
18. Cartilage	Joints	Ligaments	(Tendons)
19. Hormones	Pituitary	Pancreas	(Appendix)
20. (Thymus)	Nails	Hair	Oil glands
21. Esophagus	Pharynx	Mouth	(Trachea)
22. Thymus	Spleen	Tonsils	(Liver)

(

Fill in the missing area.

SYSTEM	ORGANS	FUNCTIONS
23. Integumentary	Skin, nails, hair, sense receptors, sweat glands, oil glands	_____
24. Skeletal	_____	Support, movement, storage of minerals, blood formation
25. Muscular	Muscles	
26. _____	Brain, spinal cord, nerves	Communication, integration, control, recognition of sensory stimuli
27. Endocrine	_____	Secretion of hormones; communication, integration, control
28. Circulatory	Heart, blood vessels	
29. Lymphatic	_____	Transportation, immune system
30. _____	Kidneys, ureters, bladder, urethra	Elimination of wastes, electrolyte balance, acid-base balance, water balance
31. Digestive	_____	Digestion of food, absorption of nutrients
32. _____	Nose, pharynx, larynx, trachea, bronchi, lungs	Exchange of gases in the lungs
33. Reproductive	_____	Survival of species; production of sex cells, fertilization, development, birth; nourishment of offspring; production of hormones

Unscramble the words.

34. RTAHE
HEART

35. IEPLNA
PINEaL

36. EENVR
NeRVE

37. SUHESOPGA
ESOPHAGUS

Take the circled letters, unscramble them, and fill in the statement.

The more thoroughly you review this chapter the less

38. NERVOUS **you will be during your test**

you have exactly ten minutes...

EXAM

▶ *If you have had difficulty with this section, review pages 46-57 and Figure 3-1.*

APPLYING WHAT YOU KNOW

39. Myrna was 15 years old and had not yet started menstruating. Her family physician decided to consult two other physicians, each of whom specialized in a different system. Specialists in the areas of _____ and _____ were consulted.

40. Brian was admitted to the hospital with second-degree and third-degree burns over 50% of his body. He was placed in isolation, so when Jenny went to visit him, she was required to wear a hospital gown and mask. Why was Brian placed in isolation? Why was Jenny required to wear special attire?

41. WORD FIND

Can you find 11 organ systems? Words may be spelled top to bottom, bottom to top, right to left, left to right, or diagonally.

```
Y R A T N E M U G E T N I R F
H N E R V O U S K I R J M G T
L Y M P H A T I C I S Y Y U I
B N X Y R O T A L U C R I C W
P E L R E O M J M S O M M P S
C C W M A N D L A T E L E K S
R R K E M L I U A A V V U K N
D K X P D J U R C J I R Q E M
C D B V C V I C C T T K W C X
X R Q Q D P H C S O I X P A Z
M F M U S Y D E V U D V Y K E
U E S E C Z G T Q D M N E K O
P Y R A N I R U C T C N E W H
N H T N D E P S I X A Q O I E
```

Circulatory   Lymphatic   Respiratory
Digestive   Muscular   Skeletal
Endocrine   Nervous   Urinary
Integumentary   Reproductive

DID YOU KNOW?

Muscles comprise 40% of your body weight. Your skeleton, however, only accounts for 18% of your body weight.

ORGAN SYSTEMS

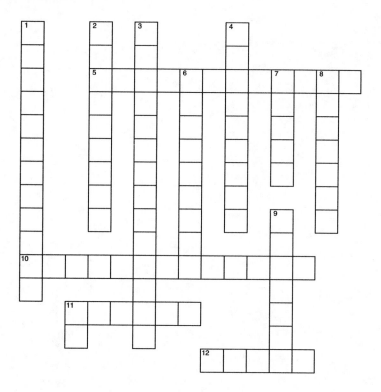

ACROSS

5. Specialized signal of nervous system (two words)
10. Skin
11. Testes and ovaries
12. Undigested residue of digestion

DOWN

1. Inflammation of the appendix
2. Vulva, penis, and scrotum
3. Heart and blood vessels
4. Subdivision of circulatory system
6. System of hormones
7. Waste product of kidneys
8. Agent that causes change in the activity of a structure
9. Chemical secretion of endocrine system
11. Gastrointestinal tract (abbreviation)

CHAPTER 4

The Integumentary System and Body Membranes

More of our time, attention, and money are spent on this system than any other one. Every time we look into a mirror we become aware of the integumentary system, as we observe our skin, hair, nails, and the appendages that give luster and comfort to this system. The discussion of the skin begins with the structure and function of the two primary layers called the epidermis and dermis. It continues with an examination of the appendages of the skin, which include the hair, receptors, nails, sebaceous glands, and sudoriferous glands. Your study of skin concludes with a review of one of the most serious and frequent threats to the skin—burns. An understanding of the integumentary system provides you with an appreciation of the danger that severe burns could pose to this system.

Membranes are thin sheetlike structures that cover, protect, anchor, or lubricate body surfaces, cavities, or organs. The two major categories are epithelial and connective. Each type is located in specific areas of the body and is vulnerable to specific disease conditions. Knowledge of the location and function of these membranes prepares you for the study of their relationship to other systems and the body as a whole.

TOPICS FOR REVIEW

Before progressing to Chapter 5, you should have an understanding of the skin and its appendages. Your review should include the classification of burns and the method used to estimate the percentage of body surface area affected by burns. A knowledge of the types of body membranes, their location, and their function is necessary to complete your study of this chapter.

CLASSIFICATION OF BODY MEMBRANES

Select the best answer.

        (a) Cutaneous   (b) Serous   (c) Mucous   (d) Synovial

_____ 1. Pleura
_____ 2. Lines joint spaces
_____ 3. Respiratory tract
_____ 4. Skin
_____ 5. Peritoneum
_____ 6. Contains no epithelium
_____ 7. Urinary tract
_____ 8. Lines body surfaces that open directly to the exterior

▶ *If you have had difficulty with this section, review pages 63-65.*

THE SKIN

Match the term on the left with the proper selection on the right.

Group A

_____ 9. Integumentary system
_____ 10. Epidermis
_____ 11. Dermis
_____ 12. Subcutaneous
_____ 13. Cutaneous membrane

A. Outermost layer of skin
B. Deeper of the two layers of skin
C. Allows for rapid absorption of injected material
D. The skin is the primary organ
E. Composed of dermis and epidermis

Group B

_____ 14. Keratin
_____ 15. Melanin
_____ 16. Stratum corneum
_____ 17. Dermal papillae
_____ 18. Cyanosis

A. Protective protein
B. Blue-gray color of skin resulting from a decrease in oxygen
C. Rows of peglike projections
D. Brown pigment
E. Outer layer of epidermis

Select the best answer.

           (a) Epidermis   (b) Dermis

_____ 19. Tightly packed epithelial cells

_____ 20. Nerves

_____ 21. Fingerprints

_____ 22. Blisters

_____ 23. Keratin

_____ 24. Connective tissue

_____ 25. Follicle

_____ 26. Sebaceous gland

_____ 27. Sweat gland

_____ 28. More cellular than other layer

▷ *If you have had difficulty with this section, review pages 65-68.*

Fill in the blanks.

29. The three most important functions of the skin are _____,
_____, and _____.

30. _____ prevents the sun's ultraviolet rays from penetrating the interior of the body.

31. The hair of a newborn infant is called _____.

32. Hair growth begins from a small, cap-shaped cluster of cells called the _____.

33. Depilatories act by dissolving the _____ in hair shafts that extend above the skin surface.

34. The _____ muscle produces "goose pimples."

35. Meissner's corpuscle is generally located rather close to the skin surface and is capable of detecting sensations of _____.

36. The most numerous, important, and widespread sweat glands in the body are the _____ sweat glands.

37. The _____ sweat glands are found primarily in the axilla and in the pigmented skin areas around the genitals.

38. _____ has been described as "nature's skin cream."

Circle the correct answer.

39. A first-degree burn (will or will not) blister.

40. A second-degree burn (will or will not) scar.

41. A third-degree burn (will or will not) have pain immediately.

42. According to the "rule of nines" the body is divided into (9 or 11) areas of 9%.

43. Destruction of the subcutaneous layer occurs in (second- or third-) degree burns.

▷ *If you have had difficulty with this section, review pages 68-75.*

Unscramble the words.

44. **PIDEEMIRS**

45. **REKTAIN**

46. **AHIR**

47. **UGONAL**

48. **DRTONIDEHYA**

Take the circled letters, unscramble them, and fill in the statement.

What Amanda's mother gave her after every date.

49.

APPLYING WHAT YOU KNOW

50. Mr. Ziven was admitted to the hospital with second-degree and third-degree burns. Both arms, the anterior trunk, the right anterior leg, and the genital region were affected by the burns. The doctor quickly estimated that _____% of Mr. Ziven's body had been burned.

51. Mrs. Tanner complained to her doctor that she had severe pain in her chest and feared that she was having a heart attack. An ECG revealed nothing unusual, but Mrs. Tanner insisted that every time she took a breath she experienced pain. What might be the cause of Mrs. Tanner's pain?

52. After investigating the scene of the crime, Officer Gorski announced that dermal papillae were found that would help solve the case. What did he mean?

53. WORD FIND

Can you find the 15 terms from this chapter in the box of letters? Words may be spelled top to bottom, bottom to top, right to left, left to right, or diagonally.

```
S  U  D  O  R  I  F  E  R  O  U  S  V  K  R
E  J  U  Q  U  E  S  T  E  C  N  O  H  Y  U
I  S  J  L  M  E  L  A  N  O  C  Y  T  E  H
R  I  M  U  E  N  O  T  I  R  E  P  H  S  B
O  M  K  N  V  A  S  T  R  L  A  N  U  G  O
T  R  G  U  F  G  A  U  C  E  G  O  V  W  D
A  E  A  L  P  R  D  I  O  N  T  F  D  R  N
L  D  V  A  D  M  L  C  P  D  H  S  L  Z  F
I  I  N  Y  N  L  R  L  A  M  E  N  I  R  E
P  P  H  Q  O  N  E  U  X  R  R  F  E  L  G
E  E  Z  F  J  U  M  Y  O  U  U  O  C  C  B
D  Z  P  E  R  J  Y  U  V  F  C  I  I  E  O
G  J  S  I  W  J  S  K  C  D  T  N  O  Z  C
C  O  S  M  Z  M  F  I  B  U  G  U  X  O  J
X  Y  P  M  E  I  W  E  C  V  S  U  G  B  I
```

Apocrine	Epidermis	Mucus
Blister	Follicle	Peritoneum
Cuticle	Lanugo	Pleurisy
Dehydration	Lunula	Serous
Depilatories	Melanocyte	Sudoriferous

DID YOU KNOW?

Because the dead cells of the epidermis are constantly being worn and washed away, we get a new outer skin layer every 27 days.

SKIN/BODY MEMBRANES

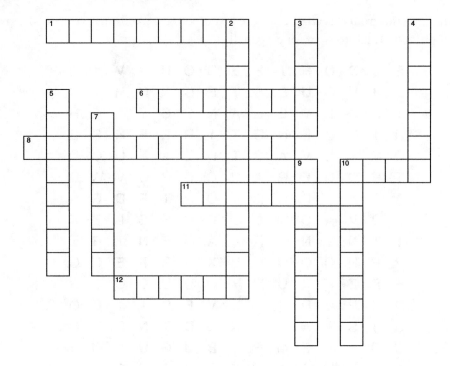

ACROSS

1. Oil gland
6. Bluish gray color of skin due to decreased oxygen
8. "Goose pimples" (two words)
10. Cutaneous
11. Inflammation of the serous membrane that lines the chest and covers the lungs
12. Deeper of the two primary skin layers

DOWN

2. Sweat gland
3. Cushionlike sac found between moving body parts
4. Brown pigment
5. Forms the lining of serous body cavities
7. Tough waterproof substance that protects body from excess fluid loss
9. Covers the surface of organs found in serous body cavities
10. Membrane that lines joint spaces

LONGITUDINAL SECTION OF THE SKIN

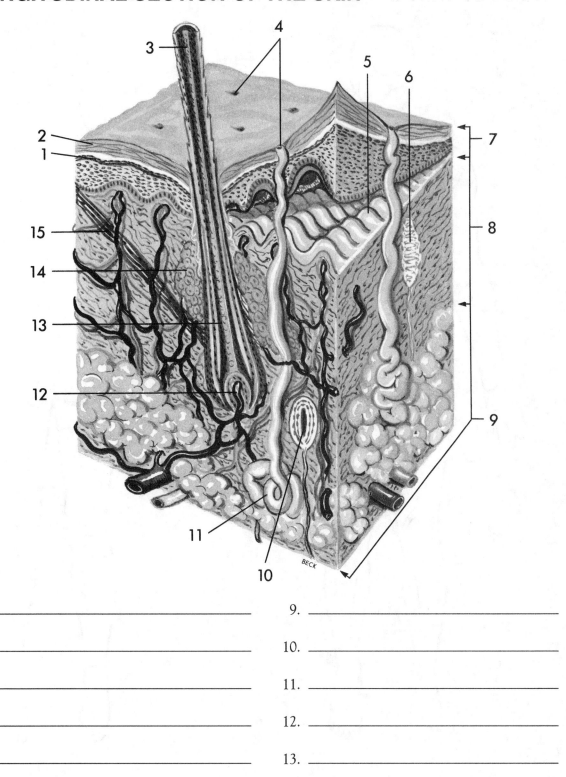

1. _____
2. _____
3. _____
4. _____
5. _____
6. _____
7. _____
8. _____
9. _____
10. _____
11. _____
12. _____
13. _____
14. _____
15. _____

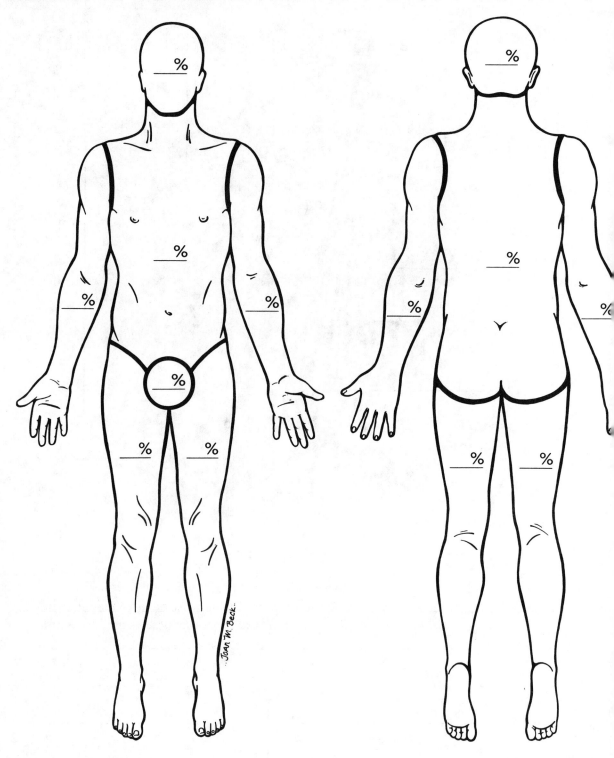

The Skeletal System

How strange we would look without our skeleton! It is the skeleton that provides us with the rigid, supportive framework that gives shape to our bodies. But this is just the beginning, since it also protects the organs beneath it, maintains homeostasis of blood calcium, produces blood cells, and assists the muscular system in providing movement for us.

After reviewing the microscopic structure of bone and cartilage, you will understand how skeletal tissues are formed, their differences, and their importance in the human body. Your microscopic investigation will make the study of this system easier as you logically progress from this view to macroscopic bone formation and growth and visualize the structure of long bones.

The skeleton is divided into two main divisions: the axial skeleton and the appendicular skeleton. All of the 206 bones of the human body may be classified into one of these two areas. Although we can divide them neatly by this system, we are still aware that subtle differences exist between a man's and a woman's skeleton. These structural differences provide us with insight to the differences in function between men and women.

Finally, three types of joints exist in the body. They are synarthrosis, amphiarthrosis, and diarthrosis. It is important to have a knowledge of these joints to understand how movement is facilitated by articulations.

TOPICS FOR REVIEW

Before progressing to Chapter 6, you should familiarize yourself with the functions of the skeletal system, the structure and function of bone and cartilage, bone formation and growth, and the types of joints found in the body. Additionally, your understanding of the skeletal system should include identification of the two major subdivisions of the skeleton, the bones found in each area, and any differences that exist between a man's and a woman's skeleton.

FUNCTIONS OF THE SKELETAL SYSTEM
TYPES OF BONES
STRUCTURE OF LONG BONES

Fill in the blanks.
1. There are _____ types of bones.
2. The _____ _____ is the hollow area inside the diaphysis of a bone.
3. A thin layer of cartilage covering each epiphysis is the _____.
4. The _____ lines the medullary cavity of long bones.
5. _____ is used to describe the process of blood cell formation.
6. Blood cell formation is a vital process carried on in _____ _____ _____.
7. The _____ is a strong fibrous membrane covering a long bone except at joint surfaces.
8. Osteoporosis occurs most frequently in _____ _____ _____.
9. Bones serve as a safety-deposit box for _____, a vital substance required for normal nerve and muscle function.
10. As muscles contract and shorten, they pull on bones and thereby _____ them.

▶ *If you have had difficulty with this section, review pages 80-82.*

MICROSCOPIC STRUCTURE OF BONE AND CARTILAGE

Match the term on the left with the proper selection on the right.
Group A
_____ 11. Trabeculae
_____ 12. Compact
_____ 13. Spongy
_____ 14. Periosteum
_____ 15. Cartilage

A. Outer covering of bone
B. Dense bone tissue
C. Fibers embedded in a firm gel
D. Needlelike threads of spongy bone
E. Ends of long bones

Group B
_____ 16. Osteocytes
_____ 17. Canaliculi
_____ 18. Lamellae
_____ 19. Chondrocytes
_____ 20. Haversian system

A. Connect lacunae
B. Cartilage cells
C. Structural unit of compact bone
D. Bone cells
E. Ring of bone

▶ *If you have had difficulty with this section, review pages 82-84.*

BONE FORMATION AND GROWTH

If the statement is true, write T in the answer blank. If the statement is false, correct the statement by circling the incorrect term and inserting the correct term in the answer blank.

_____ 21. When the skeleton forms in a baby before birth, it consists of cartilage and fibrous structures.

_____ 22. The diaphyses are the ends of the bone.

_____ 23. Bone-forming cells are known as osteoclasts.

_____ 24. It is the combined action of osteoblasts and osteoclasts that sculpts bones into their adult shapes.

_____ 25. The stresses placed on certain bones during exercise decrease the rate of bone deposition.

_____ 26. The epiphyseal plate can be seen in both external and cutaway views of an adult long bone.

_____ 27. The shaft of a long bone is known as the articulation.

_____ 28. Cartilage in the newborn becomes bone when it is replaced with calcified bone matrix deposited by osteoblasts.

_____ 29. When epiphyseal cartilage becomes bone, growth begins.

_____ 30. The epiphyseal cartilage is visible, if present, on x-ray films.

▶ *If you have had difficulty with this section, review pages 84-86.*

DIVISIONS OF SKELETON

Circle the correct choice.

31. Which one of the following is *not* a part of the axial skeleton?
    A. Scapula
    B. Cranial bones
    C. Vertebra
    D. Ribs
    E. Sternum

32. Which one of the following is *not* a cranial bone?
    A. Frontal
    B. Parietal
    C. Occipital
    D. Lacrimal
    E. Sphenoid

33. Which of the following is *not* correct?
    A. A baby is born with a straight spine.
    B. In the adult the sacral and thoracic curves are convex.
    C. The normal curves of the adult spine provide greater strength than a straight spine.
    D. A curved structure has more strength than a straight one of the same size and materials.

34. True ribs:
    A. Attach to the cartilage of other ribs
    B. Do not attach to the sternum
    C. Attach directly to the sternum without cartilage
    D. Attach directly to the sternum by means of cartilage

35. The bone that runs along the lateral side of your forearm is the:
    A. Humerus
    B. Ulna
    C. Radius
    D. Tibia

36. The shinbone is also known as the:
    A. Fibula
    B. Femur
    C. Tibia
    D. Ulna
37. The bones in the palm of the hand are called:
    A. Metatarsals
    B. Tarsals
    C. Carpals
    D. Metacarpals
38. Which one of the following is *not* a bone of the upper extremity?
    A. Radius
    B. Clavicle
    C. Humerus
    D. Ilium
39. The heel bone is known as the:
    A. Calcaneus
    B. Talus
    C. Metatarsal
    D. Phalanges
40. The mastoid process is part of which bone?
    A. Parietal
    B. Temporal
    C. Occipital
    D. Frontal
41. When a baby learns to walk, which area of the spine becomes concave?
    A. Lumbar
    B. Thoracic
    C. Cervical
    D. Coccyx
42. Which bone is the "funny" bone?
    A. Radius
    B. Ulna
    C. Humerus
    D. Carpal
43. There are how many pair of true ribs?
    A. 14
    B. 7
    C. 5
    D. 3
44. The 27 bones in the wrist and the hand allow for more:
    A. Strength
    B. Dexterity
    C. Protection
    D. Red blood cell products
45. The longest bone in the body is the
    A. Tibia
    B. Fibula
    C. Femur
    D. Humerus
46. Distally, the _____ articulates with the patella.
    A. Femur
    B. Fibula
    C. Tibia
    D. Humerus
47. These bones form the cheek bones:
    A. Mandible
    B. Palatine
    C. Maxillary
    D. Zygomatic
48. In a child, there are five of these bones. In an adult, they are fused into one:
    A. Pelvic
    B. Lumbar vertebrae
    C. Sacrum
    D. Carpals
49. The spinal cord enters the cranium through a large hole (foramen magnum) in this bone:
    A. Temporal
    B. Parietal
    C. Occipital
    D. Sphenoid

Circle the one that does not belong.

50. Cervical	Thoracic	Coxal bone	Coccyx
51. Pelvic girdle	Ankle	Wrist	Axial
52. Frontal	Occipital	Maxilla	Sphenoid
53. Scapula	Pectoral girdle	Ribs	Clavicle

54. Malleus	Vomer	Incus	Stapes
55. Ulna	Ilium	Ischium	Pubis
56. Carpal	Phalanges	Metacarpal	Ethmoid
57. Ethmoid	Parietal	Occipital	Nasal
58. Anvil	Atlas	Axis	Cervical

▷ *If you have had difficulty with this section, review pages 87-100.*

DIFFERENCES BETWEEN A MAN'S AND A WOMAN'S SKELETON

Choose the right answer.

    (a) Male   (b) Female
_____ 59. Funnel-shaped pelvis
_____ 60. Broader-shaped pelvis
_____ 61. Osteoporosis occurs more frequently
_____ 62. Larger
_____ 63. Wider pelvic inlet

▷ *If you have had difficulty with this section, review pages 100-101.*

BONE MARKINGS

From the choices given, match the bone with its identification marking. Bones may be used more than once.

A. Mastoid	H. Glenoid cavity	O. Medial malleolus
B. Pterygoid process	I. Olecranon process	P. Calcaneus
C. Foramen magnum	J. Ischium	Q. Acromion process
D. Sella turcica	K. Acetabulum	R. Frontal sinuses
E. Mental foramen	L. Symphysis pubis	S. Condyloid process
F. Conchae	M. Ilium	T. Tibial tuberosity
G. Xiphoid process	N. Greater trochanter	

_____ 64. Occipital   _____ 71. Sphenoid
_____ 65. Sternum   _____ 72. Ethmoid
_____ 66. Coxal bone   _____ 73. Scapula
_____ 67. Femur   _____ 74. Tibia
_____ 68. Ulna   _____ 75. Frontal
_____ 69. Temporal   _____ 76. Mandible
_____ 70. Tarsals

▷ *If you have had difficulty with this section, review pages 87-102.*

JOINTS (ARTICULATIONS)

Circle the correct answer.

77. Freely movable joints are (amphiarthroses or diarthroses).
78. The sutures in the skull are (synarthrotic or amphiarthrotic) joints.
79. All (diarthrotic or amphiarthrotic) joints have a joint capsule, a joint cavity, and a layer of cartilage over the ends of the two joining bones.
80. (Ligaments or tendons) grow out of periosteum and attach two bones together.
81. The (articular cartilage or epiphyseal cartilage) absorbs jolts.
82. Gliding joints are the (least movable or most movable) of the diarthrotic joints.
83. The knee is the (largest or smallest) joint.
84. Hinge joints allow motion in (2 or 4) directions.
85. The saddle joint at the base of each of our thumbs allows for greater (strength or mobility).
86. When you rotate your head, you are using a (gliding or pivot) joint.

▷ *If you have had difficulty with this section, review pages 103-106.*

Unscramble the bones.

87. ETVERRBAE

88. BPSUI

89. SCALUPA

90. IMDBALNE

91. APNHGAELS

Take the circled letters, unscramble them, and fill in the statement.

What the fat lady wore to the ball.

92.

APPLYING WHAT YOU KNOW

93. Mrs. Perine had advanced cancer of the bone. As the disease progressed, Mrs. Perine required several blood transfusions throughout her therapy. She asked the doctor one day to explain the necessity for the transfusions. What explanation might the doctor give to Mrs. Perine?

94. Dr. Kennedy, an orthopedic surgeon, called the admissions office of the hospital and advised that he would be admitting a patient in the next hour with an epiphyseal fracture. Without any other information, the patient is assigned to the pediatric ward. What prompted this assignment?

95. Mrs. Van Skiver, age 60, noticed when she went in for her physical examination that she was a half inch shorter than she was on her last visit. Dr. Veazey suggested she begin a regimen of dietary supplements of calcium, vitamin D, and a prescription for sex hormone therapy. What bone disease did Dr. Veazey suspect?

Can you find 14 terms from this chapter in the box of letters? Words may be spelled top to bottom, bottom to top, right to left, left to right, or diagonally.

```
A  R  T  I  C  U  L  A  T  I  O  N  N  U  T
M  M  L  T  N  I  N  G  U  I  H  J  N  C  G
P  R  P  E  R  I  O  S  T  E  U  M  A  N  B
H  V  P  G  U  A  T  B  M  E  O  P  R  F  G
I  N  V  R  N  O  B  O  F  S  M  H  X  E  R
A  G  J  O  A  J  P  E  T  O  E  O  H  T  Q
R  U  R  B  X  O  E  E  C  A  S  G  I  S  B
T  S  S  Y  I  M  O  Q  N  U  O  O  L  X  Q
H  J  I  E  A  B  P  U  N  Q  L  E  Q  K  S
R  I  S  I  L  U  C  I  L  A  N  A  C  X  R
O  I  N  A  M  A  S  Q  M  A  Q  K  E  C  T
S  T  S  A  L  C  O  E  T  S  O  U  I  D  G
E  T  S  Y  F  W  L  N  M  P  U  F  N  U  F
S  B  H  Q  H  L  O  U  S  A  R  X  I  T  V
R  P  M  P  A  F  M  G  X  K  D  S  L  G  A
```

Amphiarthroses	Fontanels	Osteoclasts
Articulation	Hemopoiesis	Periosteum
Axial	Lacunae	Sinus
Canaliculi	Lamella	Trabeculae
Compact	Osteoblasts	

DID YOU KNOW?

- The bones of the hands and feet make up more than half of the total 206 bones of the body.
- Approximately 25 million Americans have osteoporosis. Four out of five are women.
- The bones of the middle ear are mature at birth.

SKELETAL SYSTEM

ACROSS

4. Cartilage cells
6. Spaces in bones where osteocytes are found
8. Chest
9. Freely movable joints
11. Process of blood cell formation
13. Space inside cranial bone

DOWN

1. Joint
2. Suture joints
3. Bone absorbing cells
4. Type of bone
5. Ends of long bones
7. Covers long bone except at its joint surfaces
10. Division of skeleton
12. Bone cell

LONGITUDINAL SECTION OF LONG BONE

1. _____

2. _____

3. _____

4. _____

5. _____

6. _____

7. _____

8. _____

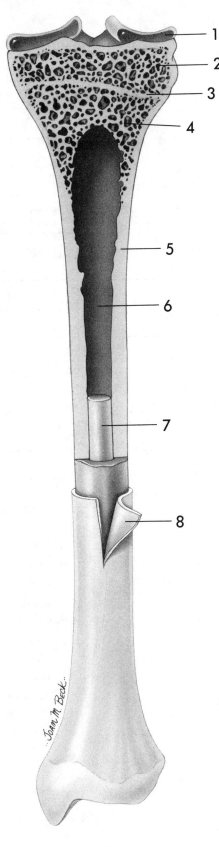

ANTERIOR VIEW OF SKELETON

1. _____

2. _____

3. _____

4. _____

5. _____

6. _____

7. _____

8. _____

9. _____

10. _____

11. _____

12. _____

13. _____

14. _____

15. _____

16. _____

17. _____

18. _____

19. _____

20. _____

21. _____

22. _____

23. _____

24. _____

25. _____

26. _____

27. _____

28. _____

29. _____

30. _____

POSTERIOR VIEW OF SKELETON

1. _____

2. _____

3. _____

4. _____

5. _____

6. _____

7. _____

8. _____

9. _____

10. _____

11. _____

12. _____

13. _____

14. _____

SKULL VIEWED FROM THE RIGHT SIDE

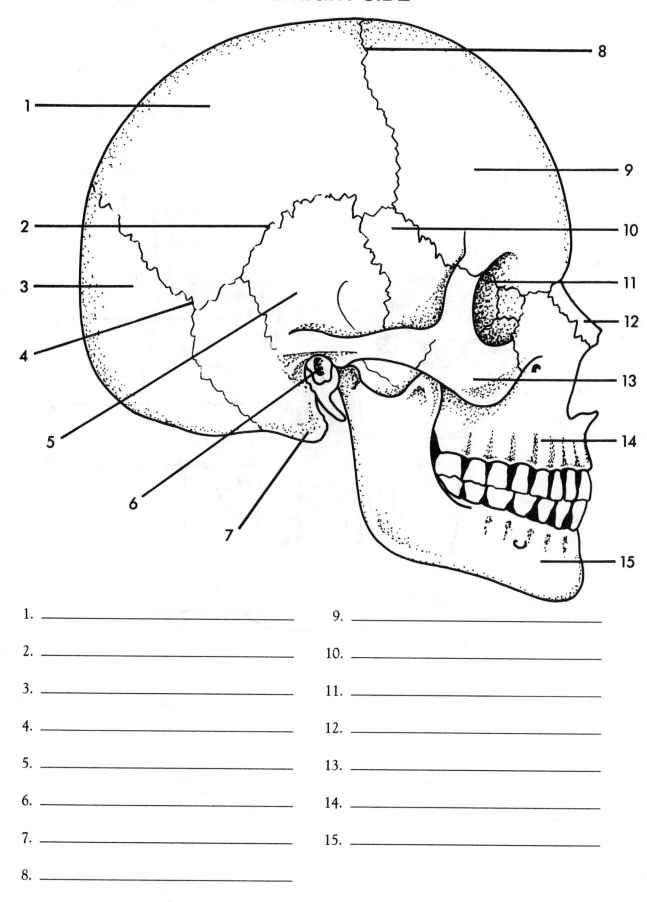

1. _____

2. _____

3. _____

4. _____

5. _____

6. _____

7. _____

8. _____

9. _____

10. _____

11. _____

12. _____

13. _____

14. _____

15. _____

SKULL VIEWED FROM THE FRONT

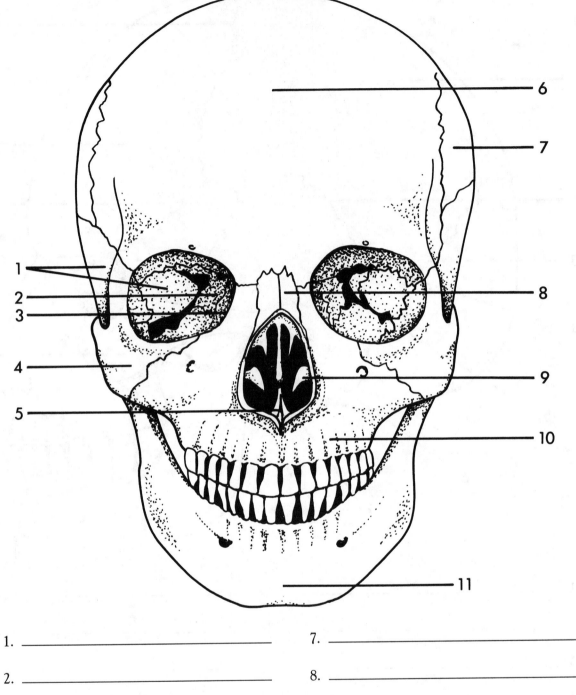

1. _____ 7. _____

2. _____ 8. _____

3. _____ 9. _____

4. _____ 10. _____

5. _____ 11. _____

6. _____

STRUCTURE OF A DIARTHROTIC JOINT

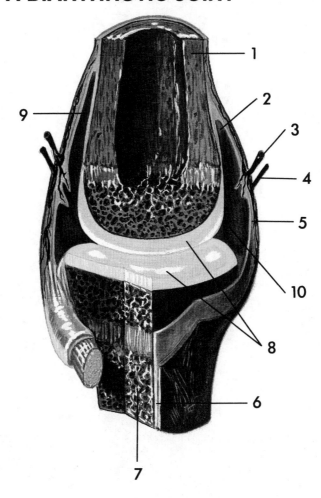

1. _____

2. _____

3. _____

4. _____

5. _____

6. _____

7. _____

8. _____

9. _____

10. _____

CHAPTER **6**   # The Muscular System

The muscular system is often referred to as the "power system," and rightfully so, because it is this system that provides the motion necessary to move the body and perform organic functions. Just as an automobile relies on the engine to provide motion, the body depends on the muscular system to perform both voluntary and involuntary types of movement. Walking, breathing, and the digestion of food are but a few examples of body functions that require the healthy performance of the muscular system.

Although this system has several functions, the primary purpose is to provide movement or power. Muscles produce power by contracting. The ability of a large muscle or muscle group to contract depends upon the ability of microscopic muscle fibers that contract within the larger muscle. An understanding of these microscopic muscle fibers will assist you as you progress in your study to the larger muscles and muscle groups.

Muscle contractions may be one of several types: isotonic, isometric, twitch, or tetanic. When skeletal or voluntary muscles contract, they provide us with a variety of motions. Flexion, extension, abduction, adduction, and rotation are examples of these movements that provide us with both strength and agility.

Muscles must be used to keep the body healthy and in good condition. Scientific evidence keeps pointing to the fact that the proper use and exercise of muscles may prolong longevity. An understanding of the structure and function of the muscular system may, therefore, add quality and quantity to our lives.

TOPICS FOR REVIEW

Before progressing to Chapter 7, you should familiarize yourself with the structure and function of the three major types of muscle tissue. Your review should include the microscopic structure of skeletal muscle tissue, how a muscle is stimulated, the major types of skeletal muscle contractions, and the skeletal muscle groups. Your study should conclude with an understanding of the types of movements produced by skeletal muscle contractions.

MUSCLE TISSUE

Select the correct term from the choices given and insert the letter(s) in the answer blank.

(a) Skeletal muscle   (b) Cardiac muscle   (c) Smooth muscle

_____ 1. Striated
_____ 2. Cells branch frequently
_____ 3. Moves food into stomach
_____ 4. Nonstriated
_____ 5. Voluntary
_____ 6. Keeps blood circulating through its vessels
_____ 7. Involuntary
_____ 8. Attaches to bone
_____ 9. Hollow internal organs
_____ 10. Maintenance of normal blood pressure

▶ *If you have had difficulty with this section, review pages 111-112.*

SKELETAL MUSCLES

Match the term on the left with the proper selection on the right.

Group A

_____ 11. Origin
_____ 12. Insertion
_____ 13. Body
_____ 14. Tendons
_____ 15. Bursae

A. The muscle unit excluding the ends
B. Attachment to the more movable bone
C. Fluid-filled sacs
D. Attachment to more stationary bone
E. Attach muscle to bones

MICROSCOPIC STRUCTURE

Group B

_____ 16. Muscle fibers
_____ 17. Actin
_____ 18. Sarcomere
_____ 19. Myosin
_____ 20. Myofilament

A. Protein that forms thick myofilaments
B. Basic functional unit of skeletal muscle
C. Protein that forms thin myofilaments
D. Microscopic threadlike structures found in skeletal muscle fibers
E. Specialized contractile cells of muscle tissue

▶ If you have had difficulty with this section, review pages 111-113.

FUNCTIONS OF SKELETAL MUSCLE

Fill in the blanks.

21. Muscles move bones by _____ on them.
22. As a rule, only the _____ bone moves.
23. The _____ bone moves toward the _____ bone.
24. Of all the muscles contracting simultaneously, the one mainly responsible for producing a particular movement is called the _____ _____ for that movement.
25. As prime movers contract, other muscles called _____ relax.
26. The biceps brachii is the prime mover during flexing, and the brachialis is its helper or _____ muscle.
27. We are able to maintain our body position because of a specialized type of skeletal muscle contraction called _____ _____.
28. _____ _____ maintains body posture by counteracting the pull of gravity.
29. A decrease in temperature, a condition known as _____, will drastically affect cellular activity and normal body function.
30. Energy required to produce a muscle contraction is obtained from _____.

▶ If you have had difficulty with this section, review pages 113-115.

FATIGUE
ROLE OF BODY SYSTEMS
MOTOR UNIT
MUSCLE STIMULUS

If the statement is true, write T on the answer blank. If the statement is false, correct the statement by circling the incorrect term and inserting the correct term in the answer blank.

_____ 31. The point of contact between the nerve ending and the muscle fiber is called a motor neuron.

_____ 32. A motor neuron together with the cells it innervates is called a motor unit.

_____ 33. If muscle cells are stimulated repeatedly without adequate periods of rest, the strength of the muscle contraction will decrease resulting in fatigue.

_____ 34. The depletion of oxygen in muscle cells during vigorous and prolonged exercise is known as fatigue.

_____ 35. An adequate stimulus will contract a muscle cell completely because of the "must" theory.

_____ 36. When oxygen supplies run low, muscle cells produce ATP and other waste products during contraction.

_____ 37. In a laboratory setting, a single muscle fiber can be isolated and subjected to stimuli of varying intensities so that it can be studied.

_____ 38. The minimal level of stimulation required to cause a fiber to contract is called the threshold stimulus.

_____ 39. Smooth muscles bring about movements by pulling on bones across movable joints.

_____ 40. A nervous system disorder that shuts off impulses to certain skeletal muscles may result in paralysis.

TYPES OF SKELETAL MUSCLE CONTRACTION

Circle the correct choice.

41. When a muscle does not shorten and no movement results, the contraction is:
    A. Isometric
    B. Isotonic
    C. Twitch
    D. Tetanic

42. Walking is an example of which type of contraction?
    A. Isometric
    B. Isotonic
    C. Twitch
    D. Tetanic

43. Pushing against a wall is an example of which type of contraction?
    A. Isotonic
    B. Isometric
    C. Twitch
    D. Tetanic

44. Endurance training is also known as:
    A. Isometrics
    B. Hypertrophy
    C. Aerobic training
    D. Strength training

45. Benefits of regular exercise include all of the following *except*:
    A. Improved lung functioning
    B. More efficient heart
    C. Less fatigue
    D. Atrophy

46. Twitch contractions can be easily seen in:
    A. Isolated muscles prepared for research
    B. A great deal of normal muscle activity
    C. During resting periods
    D. None of the above

47. Individual contractions "melt" together to produce a sustained contraction or:
    A. Twitch
    B. Tetanus
    C. Isotonic response
    D. Isometric response

48. In most cases, isotonic contraction of muscle produces movement at a/an:
    A. Insertion
    B. Origin
    C. Joint
    D. Bursa

49. Prolonged inactivity causes muscles to shrink in mass, a condition called:
    A. Hypertrophy
    B. Disuse atrophy
    C. Paralysis
    D. Muscle fatigue

50. Muscle hypertrophy can be best enhanced by a program of:
    A. Isotonic exercise
    B. Better posture
    C. High-protein diet
    D. Strength training

▶ *If you have had difficulty with this section, review pages 115-118.*

SKELETAL MUSCLE GROUPS

Choose the proper function(s) for the muscles listed below and place the letter(s) in the answer blank.

(a) Flexor   (b) Extensor   (c) Abductor

(d) Adductor   (e) Rotator   (f) Dorsiflexor or Plantar flexor

_____ 51. Deltoid

_____ 52. Tibialis anterior

_____ 53. Gastrocnemius

_____ 54. Biceps brachii

_____ 55. Gluteus medius

_____ 56. Soleus

_____ 57. Iliopsoas

_____ 58. Pectoralis major

_____ 59. Gluteus maximus

_____ 60. Triceps brachii

_____ 61. Sternocleidomastoid

_____ 62. Trapezius

_____ 63. Gracilis

▶ *If you have had difficulty with this section, review pages 119-127.*

MOVEMENTS PRODUCED BY SKELETAL MUSCLE CONTRACTIONS

Circle the correct choice.

64. A movement that makes the angle between two bones smaller is:

  A. Flexion   C. Abduction

  B. Extension   D. Adduction

65. Moving a part toward the midline is:

  A. Flexion   C. Abduction

  B. Extension   D. Adduction

66. Moving a part away from the midline is:

  A. Flexion   C. Abduction

  B. Extension   D. Adduction

67. When you move your head from side to side as in shaking your head "no" you are _____ a muscle group.

  A. Rotating   C. Supinating

  B. Pronating   D. Abducting

68. _____ occurs when you turn the palm of your hand from an anterior to posterior position.

  A. Dorsiflexion   C. Supination

  B. Plantar flexion   D. Pronation

69. Dorsiflexion refers to:

  A. Hand movements   C. Foot movements

  B. Eye movements   D. Head movements

▶ *If you have had difficulty with this section, review pages 127-129.*

APPLYING WHAT YOU KNOW

70. Casey noticed pain whenever she reached for anything in her cupboards. Her doctor told her that the small fluid-filled sacs in her shoulder were inflamed. What condition did Casey have?
71. The nurse was preparing an injection for Mrs. Tatakis. The amount to be given was 2 ml. What area of the body will the nurse most likely select for this injection?
72. Chris was playing football and pulled a band of fibrous connective tissue that attached a muscle to a bone. What is the common term for this tissue?
73. WORD FIND

Can you find the 25 muscle terms in the box of letters? Words may be spelled top to bottom, bottom to top, right to left, left to right, or diagonally.

```
G   A   S   T   R   O   C   N   E   M   I   U   S   D   U
M   S   G   I   O   N   O   I   S   N   E   T   X   E   U
U   R   N   N   T   O   M   S   R   B   S   F   D   T   T
S   U   I   S   C   I   B   O   N   T   P   N   E   A   R
C   B   R   E   U   X   V   T   O   O   E   C   L   I   A
L   Q   T   R   D   E   C   O   D   A   C   M   T   R   P
E   T   S   T   B   L   M   N   N   V   I   E   O   T   E
X   F   M   I   A   F   Y   I   E   Y   B   N   I   S   Z
S   I   A   O   V   I   E   C   T   L   S   S   D   I   I
S   Y   H   N   S   S   R   O   T   A   T   O   R   G   U
U   A   M   G   A   R   H   P   A   I   D   J   N   R   S
E   H   A   T   R   O   P   H   Y   L   N   J   T   E   B
L   N   O   H   T   D   E   U   G   I   T   A   F   N   T
O   R   I   G   I   N   O   S   N   V   S   B   Z   Y   L
S   B   V   T   O   J   C   T   R   I   C   E   P   S   S
```

Abductor	Flexion	Soleus
Atrophy	Gastrocnemius	Striated
Biceps	Hamstrings	Synergist
Bursa	Insertion	Tendon
Deltoid	Isometric	Tenosynovitis
Diaphragm	Isotonic	Trapezius
Dorsiflexion	Muscle	Triceps
Extension	Origin	
Fatigue	Rotator	

DID YOU KNOW?

If all of your muscles pulled in one direction, you would have the power to move 25 tons.

THE MUSCULAR SYSTEM

ACROSS

2. Shaking your head no
6. Muscle shrinkage
7. Toward the body's midline
9. Produces movement opposite to prime movers
12. Movement that makes joint angles larger
13. Small fluid-filled sac between tendons and bones

DOWN

1. Increase in size
3. Away from the body's midline
4. Turning your palm from an anterior to posterior position
5. Attachment to the more movable bone
7. Protein which composes myofilaments
8. Attachment to the more stationary bone
10. Assists prime movers with movement
11. Anchors muscles to bones

MUSCLES, ANTERIOR VIEW

1. _____

2. _____

3. _____

4. _____

5. _____

6. _____

7. _____

8. _____

9. _____

10. _____

11. _____

12. _____

13. _____

14. _____

15. _____

16. _____

17. _____

18. _____

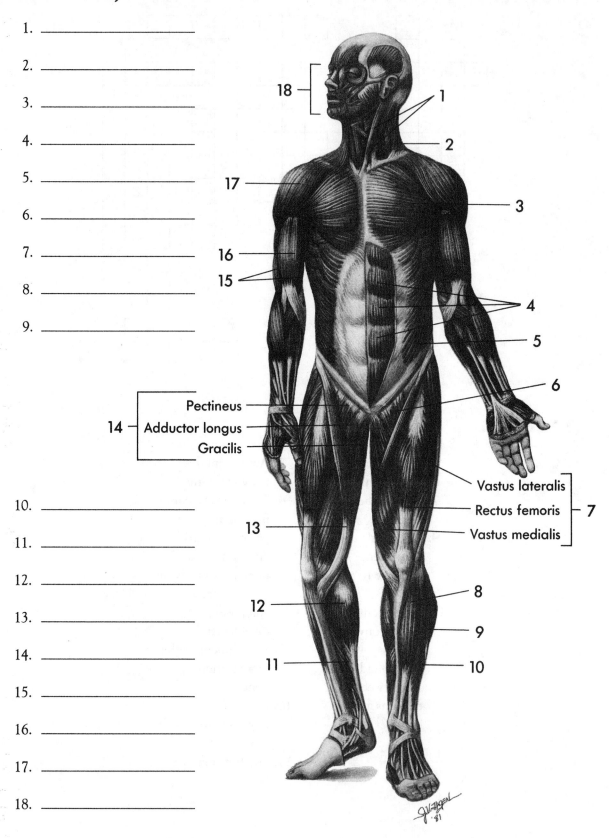

MUSCLES, POSTERIOR VIEW

1. _____

2. _____

3. _____

4. _____

5. _____

6. _____

7. _____

8. _____

9. _____

10. _____

11. _____

12. _____

13. _____

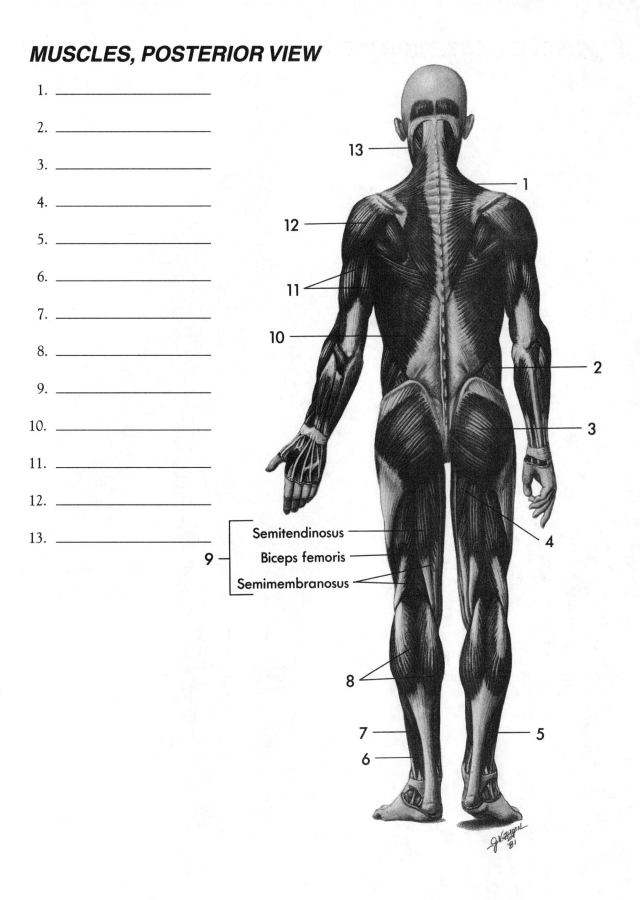

9 { Semitendinosus
    Biceps femoris
    Semimembranosus

CHAPTER 7

The Nervous System

The nervous system organizes and coordinates the millions of impulses received each day to make communication with and enjoyment of our environment possible. The functioning unit of the nervous system is the neuron. Three types of neurons, sensory, motor, and interneurons, exist and are classified according to the direction in which they transmit impulses. Nerve impulses travel over routes made up of neurons and provide the rapid communication that is necessary for maintaining life. The central nervous system is made up of the spinal cord and brain. The spinal cord provides access to and from the brain by means of ascending and descending tracts. In addition, the spinal cord functions as the primary reflex center of the body. The brain can be subdivided for easier learning into the brain stem, cerebellum, diencephalon, and cerebrum. These areas provide the extraordinary network necessary to receive, interpret, and respond to the simplest or most complex impulses.

While you concentrate on this chapter, your body is performing a multitude of functions. Fortunately for us, the beating of the heart, the digestion of food, breathing, and most of our other day-to-day processes do not require our supervision or thought. They function automatically, and the division of the nervous system that regulates these functions is known as the autonomic nervous system.

The autonomic nervous system consists of two divisions called the sympathetic system and the parasympathetic system. The sympathetic system functions as an emergency system and prepares us for "fight" or "flight." The parasympathetic system dominates control of many visceral effectors under normal everyday conditions. Together, these two divisions regulate the body's automatic functions in an effort to assist with the maintenance of homeostasis. Your understanding of this chapter will alert you to the complexity and functions of the nervous system and the "automatic pilot" of your body—the autonomic system.

TOPICS FOR REVIEW

Before progressing to Chapter 8, you should review the organs and divisions of the nervous system, the structure and function of the major types of cells in this system, the structure and function of a reflex arc, and the transmission of nerve impulses. Your study should include the anatomy and physiology of the brain and spinal cord and the nerves that extend from these two areas.

Finally, an understanding of the autonomic nervous system and the specific functions of the subdivisions of this system are necessary to complete the review of this chapter.

ORGANS AND DIVISIONS OF THE NERVOUS SYSTEM
CELLS OF THE NERVOUS SYSTEM

Match the term on the left with the proper selection on the right.

Group A

_____ 1. Sense organ

_____ 2. Central nervous system

_____ 3. Peripheral nervous system

_____ 4. Autonomic nervous system

A. Subdivision of periph-
   eral nervous system
B. Ear
C. Brain and spinal cord
D. Nerves that extend to
   the outlying parts of the
   body

Group B

_____ 5. Dendrite

_____ 6. Schwann cell

_____ 7. Motor neuron

_____ 8. Nodes of Ranvier

_____ 9. Fascicles

_____ 10. Epineurium

A. Indentations between
   adjacent Schwann cells
B. Branching projection of
   neuron
C. Also known as efferent
D. Forms myelin outside
   the CNS
E. Tough sheath that cov-
   ers the whole nerve
F. Groups of wrapped
   axons

CELLS OF NERVOUS SYSTEM

Select the best choice for the following words and insert the correct letter in the answer blank.

    (a) Neurons   (b) Neuroglia

_____ 11. Axon

_____ 12. Connective tissue

_____ 13. Astrocytes

_____ 14. Sensory

_____ 15. Conduct impulses

_____ 16. Forms the myelin sheath around central nerve fibers

_____ 17. Phagocytosis

_____ 18. Efferent

_____ 19. Multiple sclerosis

_____ 20. Neurilemma

▶ *If you have had difficulty with this section, review pages 137-141.*

REFLEX ARCS

Fill in the blanks.

21. The simplest kind of reflex arc is a _____ _____.
22. Three-neuron arcs consist of all three kinds of neurons, _____, _____, and _____.
23. Impulse conduction in a reflex arc normally starts in _____.
24. A _____ is the microscopic space that separates the axon of one neuron from the dendrites of another neuron.
25. A _____ is the response to impulse conduction over reflex arcs.
26. Contraction of a muscle that causes it to pull away from an irritating stimulus is known as the _____ _____.
27. A _____ is a group of nerve-cell bodies located in the peripheral nervous system.
28. All _____ lie entirely within the gray matter of the central nervous system.
29. In a patellar reflex, the nerve impulses that reach the quadriceps muscle (the effector) result in the classic _____ _____ response.
30. _____ _____ forms the H-shaped inner core of the spinal cord.

▷ *If you have had difficulty with this section, review pages 141-143.*

NERVE IMPULSES

Circle the correct answer.

31. Nerve impulses (do or do not) continually race along every nerve cell's surface.
32. When a stimulus acts on a neuron, it (increases or decreases) the permeability of the stimulated point of its membrane to sodium ions.
33. An inward movement of positive ions leaves a/an (lack or excess) of negative ions outside.
34. The plasma membrane of the (presynaptic neuron or postsynaptic neuron) makes up a portion of the synapse.
35. A synaptic knob is a tiny bulge at the end of the (presynaptic or postsynaptic) neuron's axon.
36. Acetylcholine is an example of a (neurotransmitter or protein molecule receptor).
37. Neurotransmitters are chemicals that allow neurons to (communicate or reproduce) with one another.
38. Neurotransmitters are distributed (randomly or specifically) into groups of neurons.
39. Catecholamines may play a role in (sleep or reproduction).
40. Endorphins and enkephalins are neurotransmitters that inhibit conduction of (fear or pain) impulses.

▷ *If you have had difficulty with this section, review pages 143-146.*

DIVISIONS OF THE BRAIN

Circle the correct choice.

41. The portion of the brain stem that joins the spinal cord to the brain is the:
    A. Pons
    B. Cerebellum
    C. Diencephalon
    D. Hypothalamus
    E. Medulla

42. Which one of the following is *not* a function of the brain stem?
    A. Conducts sensory impulses from the spinal cord to the higher centers of the brain.
    B. Conducts motor impulses from the cerebrum to the spinal cord.
    C. Controls heartbeat, respiration, and blood vessel diameter.
    D. Contains centers for speech and memory.

43. Which one of the following is *not* part of the diencephalon?
    A. Cerebrum
    B. Thalamus
    C. Pituitary gland
    D. Third ventricle gray matter

44. ADH is produced by the:
    A. Pituitary gland
    B. Medulla
    C. Mammillary bodies
    D. Third ventricle
    E. Hypothalamus

45. Which one of the following is *not* a function of the hypothalamus?
    A. It helps control the rate of heartbeat.
    B. It helps control the constriction and dilation of blood vessels.
    C. It helps control the contraction of the stomach and intestines.
    D. It produces releasing hormones that control the release of certain anterior pituitary hormones.
    E. All of the above are functions of the hypothalamus.

46. Which one of the following parts of the brain helps in the association of sensations with emotions, as well as aiding in the arousal or alerting mechanism?
    A. Pons
    B. Hypothalamus
    C. Cerebellum
    D. Thalamus
    E. None of the above is correct

47. Which of the following is *not* true of the cerebrum?
    A. Its lobes correspond to the bones that lie over them.
    B. Its grooves are called gyri.
    C. Most of its gray matter lies on the surface of the cerebrum.
    D. Its outer region is called the cerebral cortex.
    E. Its two hemispheres are connected by a structure called the corpus callosum.

48. Which one of the following is *not* a function of the cerebrum?
    A. Willed movement
    B. Consciousness
    C. Memory
    D. Conscious awareness of sensations
    E. All of the above are functions of the cerebrum

49. The area of the cerebrum responsible for the perception of sound lies in the _____ lobe.
    A. Frontal
    B. Temporal
    C. Occipital
    D. Parietal

50. Visual perception is located in the _____ lobe.
    A. Frontal
    B. Temporal
    C. Parietal
    D. Occipital
    E. None of the above is correct
51. Which one of the following is *not* a function of the cerebellum?
    A. Maintains equilibrium
    B. Helps produce smooth, coordinated movements
    C. Helps maintain normal postures
    D. Associates sensations with emotions
52. Within the interior of the cerebrum are a few islands of gray matter known as:
    A. Fissures
    B. Basal ganglia
    C. Gyri
    D. Myelin
53. A cerebrovascular accident is commonly referred to as (a):
    A. Stroke
    B. Parkinson's disease
    C. Tumor
    D. Multiple sclerosis
54. Parkinson's disease is a disease of the:
    A. Myelin
    B. Axons
    C. Neuroglia
    D. Cerebral nuclei
55. The largest section of the brain is the:
    A. Cerebellum
    B. Pons
    C. Cerebrum
    D. Midbrain

▶ *If you have had difficulty with this section, review pages 146-150 and Table 7-1.*

SPINAL CORD

If the statement is true, write T on the answer blank. If the statement is false, correct the statement by circling the incorrect term and inserting the correct term in the answer blank.

_____ 56. The spinal cord is approximately 24 to 25 inches long.
_____ 57. The spinal cord ends at the bottom of the sacrum.
_____ 58. The extension of the meninges beyond the cord is convenient for performing CAT scans without danger of injuring the spinal cord.
_____ 59. Bundles of myelinated nerve fibers—dendrites—make up the white outer columns of the spinal cord.
_____ 60. Ascending tracts conduct impulses up the cord to the brain and descending tracts conduct impulses down the cord from the brain.
_____ 61. Tracts are functional organizations in that all the axons that compose a tract serve several functions.
_____ 62. A loss of sensation caused by a spinal cord injury is called paralysis.

▶ *If you have had difficulty with this section, review pages 150-154 and 156.*

COVERINGS AND FLUID SPACES OF BRAIN AND SPINAL CORD

Circle the one that does not belong.

63. Meninges   Pia mater   Ventricles   Dura mater
64. Arachnoid   Middle layer   CSF   Cobweblike
65. CSF   Ventricles   Subarachnoid space   Pia mater
66. Tough   Outer layer   Dura mater   Choroid plexus
67. Brain tumor   Subarachnoid space   CSF   Fourth lumbar vertebra

CRANIAL NERVES

68. Fill in the missing areas on the chart below.

NERVE		CONDUCT IMPULSES	FUNCTION
I	_____	From nose to brain	Sense of smell
II	Optic	From eye to brain	_____
III	Oculomotor	_____	Eye movements
IV	_____	From brain to external eye muscles	Eye movements
V	Trigeminal	From skin and mucous membrane of head and from teeth to brain; also from brain to chewing muscles	_____ _____
VI	Abducens	_____	Turning eyes outward
VII	Facial	From taste buds of tongue to brain; from brain to face muscles	_____ _____
VIII	_____	From ear to brain	Hearing; sense of balance
IX	Glossopharyngeal	_____ _____	Sensations of throat, taste, swallowing movements, secretion of saliva
X	_____	From throat, larynx, and organs in thoracic and abdominal cavities to brain; also from brain to muscles of throat and to organs in thoracic and abdominal cavities	Sensations of throat, larynx, and of thoracic and abdominal organs; swallowing, voice production, slowing of heartbeat, acceleration of peristalsis (gut movements)
XI	Accessory	From brain to certain shoulder and neck muscles	_____
XII	_____	From brain to muscles of tongue	Tongue movements

▶ *If you have had difficulty with this section, review pages 154-159 and Table 7-2.*

CRANIAL NERVES
SPINAL NERVES

Select the best choice for the following words and insert the correct letter in the answer blank.

        (a) Cranial nerves   (b) Spinal nerves

_____ 69. 12 pair

_____ 70. Dermatome

_____ 71. Vagus

_____ 72. Shingles

_____ 73. 31 pair

_____ 74. Optic

_____ 75. C1

_____ 76. Plexus

▶ *If you have had difficulty with this section, review pages 156-159 and 166.*

AUTONOMIC NERVOUS SYSTEM

Match the term on the left with the proper selection on the right.

_____ 77. Autonomic nervous system

_____ 78. Autonomic neurons

_____ 79. Preganglionic neurons

_____ 80. Visceral effectors

_____ 81. Sympathetic system

_____ 82. Somatic nervous system

A. Division of ANS

B. Tissues to which autonomic neurons conduct
impulses

C. Voluntary actions

D. Regulates body's involuntary functions

E. Motor neurons that
make up the ANS

F. Conduct impulses
between the spinal cord
and a ganglion

SYMPATHETIC NERVOUS SYSTEM
PARASYMPATHETIC NERVOUS SYSTEM

Circle the correct choice.

83. Dendrites and cell bodies of sympathetic preganglionic neurons are located in the:

    A. Brain stem and sacral portion of the spinal cord

    B. Sympathetic ganglia

    C. Gray matter of the thoracic and upper lumbar segments of the spinal cord

    D. Ganglia close to effectors

84. Which of the following is *not* correct?
    A. Sympathetic preganglionic neurons have their cell bodies located in the lateral gray column of certain parts of the spinal cord.
    B. Sympathetic preganglionic axons pass along the dorsal root of certain spinal nerves.
    C. There are synapses within sympathetic ganglia.
    D. Sympathetic responses are usually widespread, involving many organs.
85. Another name for the parasympathetic nervous system is:
    A. Thoracolumbar
    B. Craniosacral
    C. Visceral
    D. ANS
    E. Cholinergic
86. Which statement is *not* correct?
    A. Sympathetic postganglionic neurons have their dendrites and cell bodies in sympathetic ganglia or collateral ganglia.
    B. Sympathetic ganglions are located in front of and at each side of the spinal column.
    C. Separate autonomic nerves distribute many sympathetic postganglionic axons to various internal organs.
    D. Very few sympathetic preganglionic axons synapse with postganglionic neurons.
87. Sympathetic stimulation usually results in:
    A. Response by numerous organs
    B. Response by only one organ
    C. Increased peristalsis
    D. Constriction of pupils
88. Parasympathetic stimulation frequently results in:
    A. Response by only one organ
    B. Responses by numerous organs
    C. The fight or flight syndrome
    D. Increased heartbeat

Choose the correct response and insert the letter in the answer blanks.

       (a) Sympathetic control   (b) Parasympathetic control

_____ 89. Constricts pupils
_____ 90. "Goose pimples"
_____ 91. Increases sweat secretion
_____ 92. Increases secretion of digestive juices
_____ 93. Constricts blood vessels

_____ 94. Slows heartbeat
_____ 95. Relaxes bladder
_____ 96. Increases epinephrine secretion
_____ 97. Increases peristalsis
_____ 98. Stimulates lens for near vision

▶ *If you have had difficulty with this section, review pages 159-164.*

AUTONOMIC NEUROTRANSMITTERS
AUTONOMIC NERVOUS SYSTEM AS A WHOLE

Fill in the blanks.

99. Sympathetic preganglionic axons release the neurotransmitter _____.
100. Axons that release norepinephrine are classified as _____.

101. Axons that release acetylcholine are classified as _____.
102. The function of the autonomic nervous system is to regulate the body's involuntary functions in ways that maintain or restore _____.
103. Your _____ _____ is determined by the combined forces of the sympathetic and parasympathetic nervous system.
104. According to some physiologists, meditation leads to _____ sympathetic activity and changes opposite to those of the fight or flight syndrome.

▷ *If you have had difficulty with this section, review pages 164-166.*

Unscramble the words.

105. **RONNESU**

106. **APSYENS**

107. **CIATUNOMO**

108. **SHTOMO**   **ULMSEC**

Take the circled letters, unscramble them, and fill in the statement.

What the man hoped the IRS agent would be during his audit.

109.

APPLYING WHAT YOU KNOW

110. Mr. Hemstreet suffered a cerebrovascular accident and it was determined that the damage affected the left side of his cerebrum. On which side of his body will he most likely notice any paralysis?
111. Baby Dania was born with an excessive accumulation of cerebrospinal fluid in the ventricles. A catheter was placed in the ventricle and the fluid was drained by means of a shunt into the circulatory bloodstream. What condition does this medical history describe?

112. Mrs. Muhlenkamp looked out her window to see a man trapped under the wheel of a car. Although slightly built, Mrs. Muhlenkamp rushed to the car, lifted it, and saved the man underneath the wheel. What division of the autonomic nervous system made this seemingly impossible task possible?

113. Lynn's heart raced and her palms became clammy as she watched the monster at the local theater. When the movie was over, however, she told her friends that she was not afraid at all. She appeared to be as calm as before the movie. What division of the autonomic nervous system made this possible?

114. Bill is going to his boss for his annual evaluation. He is planning to ask for a raise and hopes the evaluation will be good. Which subdivision of the autonomic nervous system will be active during this conference? Should he have a large meal before his appointment? Support your answer with facts from the chapter.

115. WORD FIND

Can you find the 14 terms from this chapter in the box of letters? Words may be spelled top to bottom, bottom to top, right to left, left to right, or diagonally.

```
M C C D Q S Y N A P S E Q G O
E N A Q D W H N W E J N A L W
S R O T P E C E R Y Z N I S M
I D S X E K N O K X G G C Y Y
J O T R A C T D F L O N E N A
R P W E K O H X I D A L H A M
F A L O N A Y O E I I I X P C
C M Z I F D N N L N G Z U T Z
S I N V C G D G Z A N A K I P
G N A T Z R O W V A M Y T C O
K E N D O R P H I N S I L C X
C P Q G C J X F J D Q S N L F
X U L I H I G Q A N S W O E U
A I M H Q E X K D B W Y T F S
A A D A X O C K G B F H B T K
```

Axon	Glia	Serotonin
Catecholamines	Microglia	Synapse
Dopamine	Myelin	Synaptic cleft
Endorphins	Oligodendroglia	Tract
Ganglion	Receptors	

DID YOU KNOW?

Although all pain is felt and interpreted in the brain, it has no pain sensation itself—even when cut!

THE NERVOUS SYSTEM

ACROSS

4. Bundle of axons located within the CNS

8. Transmits impulses toward the cell body

9. Neurons that conduct impulses from a ganglion

10. Astrocytes

11. Pia mater

13. Nerve cells

14. Transmits impulses away from the cell body

15. Peripheral nervous system (abbreviation)

DOWN

1. Neuroglia

2. Peripheral beginning of a sensory neuron's dendrite

3. Two neuron arc (two words)

5. Neurotransmitter

6. Cluster of nerve cell bodies outside the central nervous system

7. Area of brain stem

11. Fatty substance found around some nerve fibers

12. Where impulses are transmitted from one neuron to another

NEURON

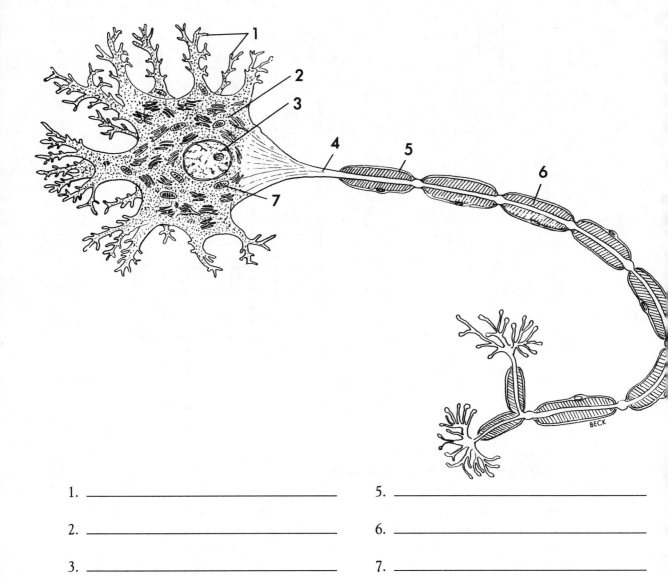

1. _____   5. _____

2. _____   6. _____

3. _____   7. _____

4. _____

CRANIAL NERVES

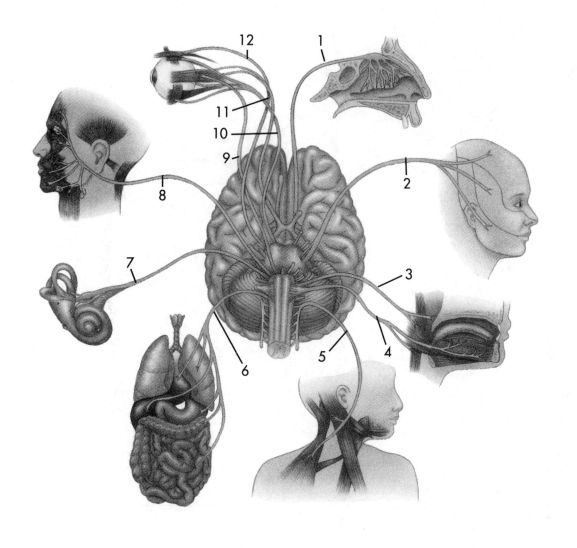

1. _____   7. _____

2. _____   8. _____

3. _____   9. _____

4. _____   10. _____

5. _____   11. _____

6. _____   12. _____

NEURAL PATHWAY INVOLVED IN THE PATELLAR REFLEX

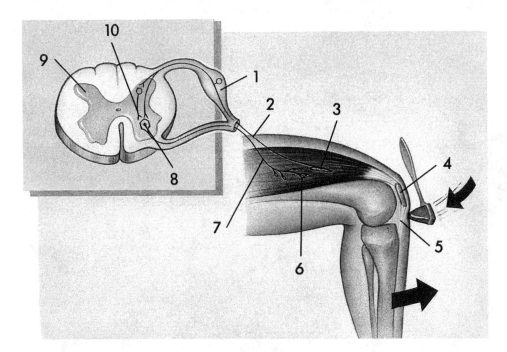

1. _____   6. _____

2. _____   7. _____

3. _____   8. _____

4. _____   9. _____

5. _____   10. _____

THE CEREBRUM

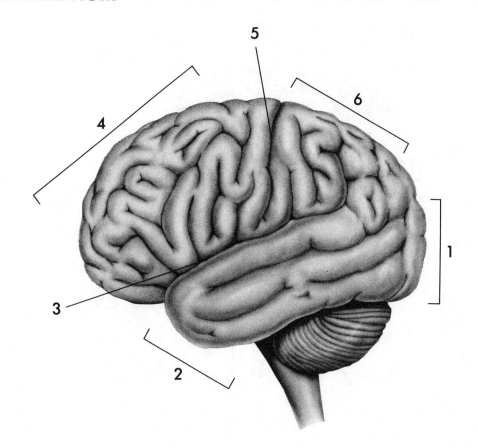

1. _____

2. _____

3. _____

4. _____

5. _____

6. _____

SAGITTAL SECTION OF THE CENTRAL NERVOUS SYSTEM

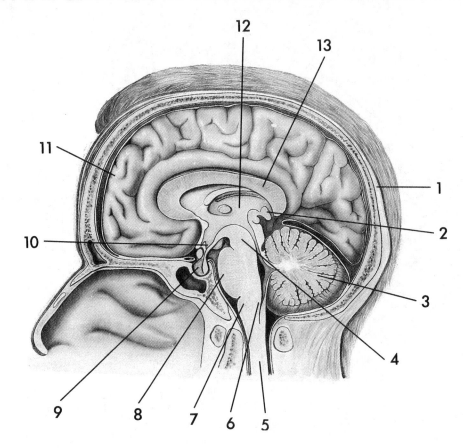

1. _____
2. _____
3. _____
4. _____
5. _____
6. _____
7. _____

8. _____
9. _____
10. _____
11. _____
12. _____
13. _____

NEURON PATHWAYS

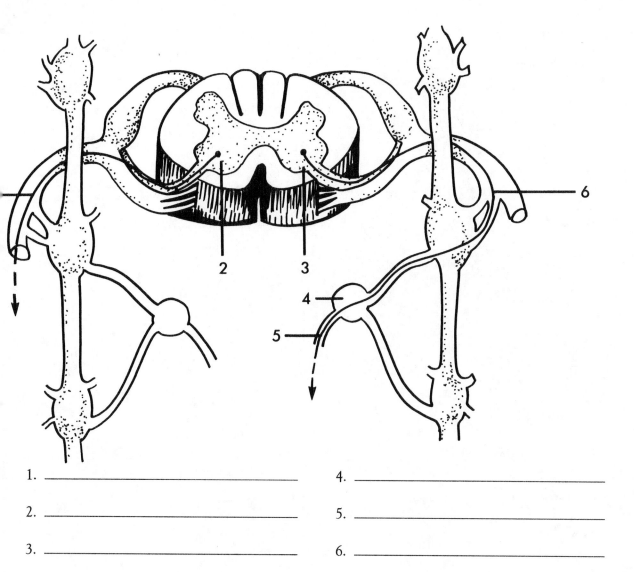

1. _____   4. _____

2. _____   5. _____

3. _____   6. _____

CHAPTER **8**

The Senses

Consider this scene for a moment. You are walking along a beautiful beach watching the sunset. You notice the various hues and are amazed at the multitude of shades that cover the sky. The waves are indeed melodious as they splash along the shore and you wiggle your feet with delight as you sense the warm, soft sand trickling between your toes. You sip on a soda and then inhale the fresh salt air as you continue your stroll along the shore. It is a memorable scene, but one that would not be possible without the assistance of your sense organs. The sense organs pick up messages that are sent over nerve pathways to specialized areas in the brain for interpretation. They make communication with and enjoyment of the environment possible. The visual, auditory, tactile, olfactory, and gustatory sense organs not only protect us from danger but also add an important dimension to our daily pleasures of life.

Your study of this chapter will give you an understanding of another one of the systems necessary for homeostasis and survival.

TOPICS FOR REVIEW

Before progressing to Chapter 9, you should review the classification of sense organs and the process for converting a stimulus into a sensation. Your study should also include an understanding of the special sense organs and the general sense organs.

CLASSIFICATION OF SENSE ORGANS
CONVERTING A STIMULUS INTO A SENSATION
GENERAL SENSE ORGANS

Match the term on the left with the proper selection on the right.

_____ 1. Special sense organ
_____ 2. General sense organ
_____ 3. Nose
_____ 4. Krause's end-bulbs
_____ 5. Taste buds

A. Olfactory cells
B. Meissner's corpuscles
C. Chemoreceptor
D. Eye
E. Touch

▶ *If you have had difficulty with this section, review pages 173-176.*

SPECIAL SENSE ORGANS

Eye

Circle the correct choice.

6. The "white" of the eye is more commonly called the:
   A. Choroid
   B. Cornea
   C. Sclera
   D. Retina
   E. None of the above is correct

7. The "colored" part of the eye is known as the:
   A. Retina
   B. Cornea
   C. Pupil
   D. Sclera
   E. Iris

8. The transparent portion of the sclera, referred to as the "window" of the eye, is the:
   A. Retina
   B. Cornea
   C. Pupil
   D. Iris

9. The mucous membrane that covers the front of the eye is called the:
   A. Cornea
   B. Choroid
   C. Conjunctiva
   D. Ciliary body
   E. None of the above is correct

10. The structure that can contract or dilate to allow more or less light to enter the eye is the:
    A. Lens
    B. Choroid
    C. Retina
    D. Cornea
    E. Iris

11. When the eye is looking at objects far in the distance, the lens is _____ and the ciliary muscle is _____.
    A. Rounded, contracted
    B. Rounded, relaxed
    C. Slightly rounded, contracted
    D. Slightly curved, relaxed
    E. None of the above is correct

12. The lens of the eye is held in place by the:
    A. Ciliary muscle   C. Vitreous humor
    B. Aqueous humor   D. Cornea
13. When the lens loses its elasticity and can no longer bring near objects into focus, the condition is known as:
    A. Glaucoma   C. Astigmatism
    B. Presbyopia   D. Strabismus
14. The fluid in front of the lens that is constantly being formed, drained, and replaced in the anterior cavity is the:
    A. Vitreous humor   C. Aqueous humor
    B. Protoplasm   D. Conjunctiva
15. If drainage of the aqueous humor is blocked, the internal pressure within the eye will increase and a condition known as _____ could occur.
    A. Presbyopia   C. Color blindness
    B. Glaucoma   D. Cataracts
16. The rods and cones are the photoreceptor cells and are located on the:
    A. Sclera   C. Choroid
    B. Cornea   D. Retina
17. The area that contains the greatest concentration of cones on the retina is the:
    A. Fovea centralis   C. Ciliary body
    B. Retinal artery   D. Optic disc
18. If our eyes are abnormally elongated, the image focuses in front of the retina and a condition known as _____ occurs.
    A. Hyperopia   C. Night blindness
    B. Cataracts   D. Myopia

▶ *If you have had difficulty with this section, review pages 176-179.*

Ear

Select the best answer from the choices given and insert the letter in the answer blank.

    (a) External ear   (b) Middle ear   (c) Inner ear

_____ 19. Malleus
_____ 20. Perilymph
_____ 21. Incus
_____ 22. Ceruminous glands
_____ 23. Cochlea
_____ 24. Auditory canal
_____ 25. Semicircular canals
_____ 26. Stapes
_____ 27. Eustachian tube
_____ 28. Organ of Corti

Fill in the blanks.

29. The external ear has two parts: the _____ and the

_____.

30. Another name for the tympanic membrane is the _____.

31. The bones of the middle ear are referred to, collectively, as _____.

32. The stapes presses against a membrane that covers a small opening, the _____.

33. A middle ear infection is called _____ _____.

34. The _____ is located adjacent to the oval window between the semicircular canals and the cochlea.

35. Located within the semicircular canals and the vestibule are _____ for balance and equilibrium.

36. The sensory cells in the _____ _____ are stimulated when movement of the head causes the endolymph to move.

▶ *If you have had difficulty with this section, review pages 180-184.*

TASTE RECEPTORS
SMELL RECEPTORS

Circle the correct answer.

37. Structures known as (papillae or olfactory cells) are found on the tongue.

38. Nerve impulses generated by stimulation of taste buds travel primarily through two (cranial or spinal) nerves.

39. To be detected by olfactory receptors, chemicals must be dissolved in the watery (mucus or plasma) that lines the nasal cavity.

40. The pathways taken by olfactory nerve impulses and the areas where these impulses are interpreted are closely associated with areas of the brain important in (hearing or memory).

41. (Chemoreceptor or mechanoreceptor) is the term used to describe the type of receptors that generate nervous impulses resulting in the sense of taste or smell.

▶ *If you have had difficulty with this section, review pages 184-185.*

APPLYING WHAT YOU KNOW

42. Mr. Nay was an avid swimmer and competed regularly in his age group. He had to withdraw from the last competition due to an infection of his ear. Antibiotics and analgesics were prescribed by the doctor. What is the medical term for his condition?

43. Mrs. Metheny loved the out-of-doors and spent a great deal of her spare time basking in the sun on the beach. Her physician suggested that she begin wearing sunglasses regularly when he noticed milky spots beginning to appear on Mrs. Metheny's lenses. What condition was Mrs. Metheny's physician trying to prevent from occurring?

44. Amanda repeatedly became ill with throat infections during her first few years of school. Lately, however, she has noticed that whenever she has a throat infection, her ears become very sore also. What might be the cause of this additional problem?

45. Julius was hit in the nose with a baseball during practice. His sense of smell was temporarily gone. What nerve receptors were damaged during the injury?

46. WORD FIND

Can you find the 19 terms from this chapter in the box of letters? Words may be spelled top to bottom, bottom to top, right to left, left to right, or diagonally.

```
M  E  C  H  A  N  O  R  E  C  E  P  T  O  R
H  R  A  T  B  Q  I  R  T  B  N  H  M  A  E
P  F  T  Y  R  O  T  C  A  F  L  O  X  R  C
G  U  A  Y  I  N  A  I  H  C  A  T  S  U  E
G  P  R  A  C  E  R  U  M  E  N  O  P  X  P
I  E  A  I  P  O  Y  B  S  E  R  P  K  D  T
W  L  C  P  L  Y  N  E  Y  K  E  I  T  O  O
C  Q  T  O  I  B  R  J  B  S  F  G  L  M  R
Z  U  S  R  C  L  E  O  U  J  R  M  N  N  S
F  D  M  E  O  H  L  Q  T  N  A  E  Y  I  S
H  I  D  P  N  D  L  A  M  A  C  N  X  S  Z
A  M  L  Y  E  S  S  E  E  D  T  T  M  Z  H
D  D  M  H  S  F  E  X  A  Y  I  S  I  I  A
J  C  G  N  J  T  I  S  L  P  O  G  U  V  J
P  H  G  Y  A  K  H  S  U  C  N  I  G  G  A
```

Cataracts	Gustatory	Presbyopia
Cerumen	Hyperopia	Receptors
Cochlea	Incus	Refraction
Cones	Mechanoreceptor	Rods
Conjunctiva	Olfactory	Senses
Eustachian	Papillae	
Eye	Photopigment	

THE SENSES

ACROSS

2. Bones of the middle ear
4. Located in anterior cavity in front of lens (two words)
5. External ear
6. Transparent body behind pupil
8. Front part of this coat is the ciliary muscle and iris
10. Membranous labyrinth filled with this fluid

DOWN

1. Located in posterior cavity (two words)
3. Organ of Corti located here
7. White of the eye
9. Innermost layer of the eye
11. Hole in center of iris

EYE

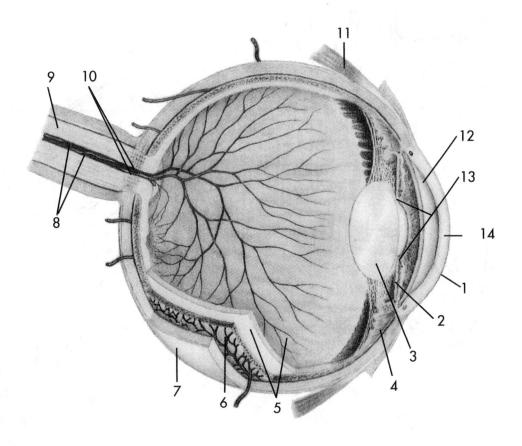

1. _____

2. _____

3. _____

4. _____

5. _____

6. _____

7. _____

8. _____

9. _____

10. _____

11. _____

12. _____

13. _____

14. _____

EAR

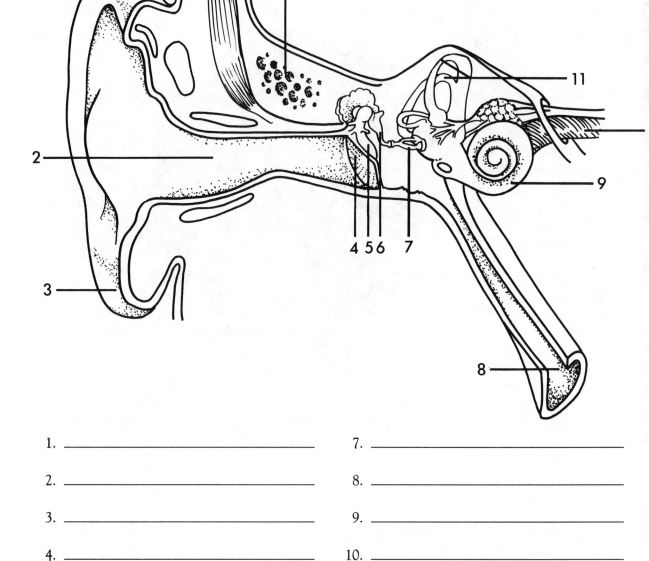

1. _____   7. _____

2. _____   8. _____

3. _____   9. _____

4. _____   10. _____

5. _____   11. _____

6. _____

The Endocrine System

The endocrine system has often been compared to a fine concert symphony. When all instruments are playing properly, the sound is melodious. If one instrument plays too loud or too soft, however, it affects the overall quality of the entire performance.

The endocrine system is a ductless system that releases hormones into the bloodstream to help regulate body functions. The pituitary gland may be considered the conductor of the orchestra, as it stimulates many of the endocrine glands to secrete their powerful hormones. All hormones, whether stimulated in this manner or by other control mechanisms, are interdependent. A change in the level of one hormone may affect the level of many other hormones.

In addition to the endocrine glands, prostaglandins, or "tissue hormones," are powerful substances similar to hormones that have been found in a variety of body tissues. These hormones are often produced in a tissue and diffuse only a short distance to act on cells within that area. Prostaglandins influence respiration, blood pressure, gastrointestinal secretions, and the reproductive system, and may some day play an important role in the treatment of diseases such as hypertension, asthma, and ulcers.

The endocrine system is a system of communication and control. It differs from the nervous system in that hormones provide a slower, longer lasting effect than do nerve stimuli and responses. Your understanding of the "system of hormones" will alert you to the mechanism of our emotions, response to stress, growth, chemical balances, and many other body functions.

TOPICS FOR REVIEW

Before progressing to Chapter 10, you should be able to identify and locate the primary endocrine glands of the body. Your understanding should include the hormones that are produced by these glands and the method by which these secretions are regulated. Your study will conclude with the pathological conditions that result from the malfunctioning of this system.

MECHANISMS OF HORMONE ACTION
REGULATION OF HORMONE SECRETION
PROSTAGLANDINS

Match the term on the left with the proper selection on the right.

Group A

_____ 1. Pituitary
_____ 2. Parathyroids
_____ 3. Adrenals
_____ 4. Ovaries
_____ 5. Thymus

A. Pelvic cavity
B. Mediastinum
C. Neck
D. Cranial cavity
E. Abdominal cavity

Group B

_____ 6. Negative feedback
_____ 7. Tissue hormones
_____ 8. Second messenger hypothesis
_____ 9. Exocrine glands
_____ 10. Target organ cells

A. Explanation for hormone organ recognition
B. Respond to a particular hormone
C. Prostaglandins
D. Discharge secretions into ducts
E. Specialized homeostatic mechanism that regulates release of hormones

Fill in the blanks.

The (11) _____ _____ hypothesis is a theory that attempts

to explain why hormones cause specific effects in target organs but do not (12) _____

or act on other organs of the body. Protein hormones serve as (13) _____

_____, providing communication between endocrine glands and

(14) _____ _____. The second messenger (15) _____

_____ provides communication within a hormone's

(16) _____ _____. (17) _____

_____ disrupts the normal negative feedback control of hormones throughout the

body, and may result in tissue damage, sterility, mental imbalance, and a host of life-threatening meta-

bolic problems.

▶ *If you have had difficulty with this section, review pages 191-196.*

PITUITARY GLAND
HYPOTHALAMUS

Circle the correct choice.

18. The pituitary gland lies in the _____ bone.
    - A. Ethmoid
    - B. Sphenoid
    - C. Temporal
    - D. Frontal
    - E. Occipital

19. Which one of the following structures would *not* be stimulated by a tropic hormone from the anterior pituitary?
    - A. Ovaries
    - B. Testes
    - C. Thyroid
    - D. Adrenals
    - E. Uterus

20. Which one of the following is *not* a function of FSH?
    - A. Stimulates the growth of follicles
    - B. Stimulates the production of estrogens
    - C. Stimulates the growth of seminiferous tubules
    - D. Stimulates the interstitial cells of the testes

21. Which one of the following is *not* a function of LH?
    - A. Stimulates maturation of a developing follicle
    - B. Stimulates the production of estrogens
    - C. Stimulates the formation of a corpus luteum
    - D. Stimulates sperm cells to mature in the male
    - E. Causes ovulation to occur

22. Which one of the following is *not* a function of GH?
    - A. Increases glucose catabolism
    - B. Increases fat catabolism
    - C. Speeds up the movement of amino acids into cells from the bloodstream
    - D. All of the above are functions of GH

23. Which one of the following hormones is *not* released by the anterior pituitary gland?
    - A. ACT
    - D. FSH
    - B. TSH
    - E. LH
    - C. ADH

24. Which one of the following is *not* a function of prolactin?
    - A. Stimulates breast development during pregnancy
    - B. Stimulates milk secretion after delivery
    - C. Causes the release of milk from glandular cells of the breast
    - D. All of the above are functions of prolactin

25. The anterior pituitary gland:
    - A. Secretes eight major hormones
    - B. Secretes tropic hormones that stimulate other endocrine glands to grow and secrete
    - C. Secretes ADH
    - D. Secretes oxytocin

26. TSH acts on the:
    - A. Thyroid
    - B. Thymus
    - C. Pineal
    - D. Testes

27. ACTH stimulates the:
   A. Adrenal cortex   C. Hypothalamus
   B. Adrenal medulla   D. Ovaries
28. Which hormone is secreted by the posterior pituitary gland?
   A. MSH   C. GH
   B. LH   D. ADH
29. ADH serves the body by:
   A. Initiating labor
   B. Accelerating water reabsorption from urine into the blood
   C. Stimulating the pineal gland
   D. Regulating the calcium/phosphorus levels in the blood
30. The disease caused by hyposecretion of the ADH is:
   A. Diabetes insipidus
   B. Diabetes mellitus
   C. Acromegaly
   D. Myxedema
31. The actual production of ADH and oxytocin takes place in which area?
   A. Anterior pituitary
   B. Posterior pituitary
   C. Hypothalamus
   D. Pineal
32. Inhibiting hormones are produced by the:
   A. Anterior pituitary
   B. Posterior pituitary
   C. Hypothalamus
   D. Pineal

Select the best answer from the choices given and insert the letter in the answer blank.

      (a) Anterior pituitary   (b) Posterior pituitary   (c) Hypothalamus

_____ 33. Adenohypophysis
_____ 34. Neurohypophysis
_____ 35. Induced labor
_____ 36. Appetite
_____ 37. Acromegaly
_____ 38. Body temperature
_____ 39. Sex hormones
_____ 40. Tropic hormones
_____ 41. Gigantism
_____ 42. Releasing hormones

▶ *If you have had difficulty with this section, review pages 196-199.*

THYROID GLAND
PARATHYROID GLANDS

Circle the correct answer.

43. The thyroid gland lies (above or below) the larynx.
44. The thyroid gland secretes (calcitonin or glucagon).
45. For thyroxine to be produced in adequate amounts, the diet must contain sufficient (calcium or iodine).
46. Most endocrine glands (do or do not) store their hormones.
47. Colloid is a storage medium for the (thyroid hormone or parathyroid hormone).
48. Calcitonin (increases or decreases) the concentration of calcium in the blood.
49. Simple goiter results from (hyperthyroidism or hypothyroidism).
50. Hyposecretion of thyroid hormones during the formative years leads to (cretinism or myxedema).
51. The parathyroid glands secrete the hormone (PTH or PTA).
52. Parathyroid hormone tends to (increase or decrease) the concentration of calcium in the blood.

▶ *If you have had difficulty with this section, review pages 200-203.*

ADRENAL GLANDS

Fill in the blanks.

53. The adrenal gland is actually two separate endocrine glands, the _____ _____ and the _____ - _____.
54. Hormones secreted by the adrenal cortex are known as _____.
55. The outer zone of the adrenal cortex, the zona glomerulosa, secretes _____.
56. The middle zone, the zona fasciculata, secretes _____.
57. The innermost zone, the zona reticularis, secretes _____ _____.
58. Glucocorticoids act in several ways to increase _____.
59. Glucocorticoids also play an essential part in maintaining _____ _____.
60. The adrenal medulla secretes the hormones _____ and _____.
61. The adrenal medulla may help the body resist _____.
62. The term _____ _____ _____ is often used to describe how the body mobilizes a number of different defense mechanisms when threatened by harmful stimuli.

Select the best response from the choices given and insert the letter in the answer blank.

        (a) Adrenal cortex   (b) Adrenal medulla

_____ 63. Addison's disease   _____ 67. Fight or flight syndrome

_____ 64. Anti-immunity   _____ 68. Aldosterone

_____ 65. Adrenaline   _____ 69. Androgens

_____ 66. Cushing's syndrome

▶ *If you have had difficulty with this section, review pages 203-207.*

PANCREATIC ISLETS
SEX GLANDS
THYMUS
PLACENTA
PINEAL GLAND

Circle the term that does not belong.

70. Alpha cells	Glucagon	Beta cells	Glycogenolysis
71. Insulin	Glucagon	Beta cells	Diabetes mellitus
72. Estrogens	Progesterone	Corpus luteum	Thymosin
73. Chorion	Interstitial cells	Testosterone	Semen
74. Immune system	Mediastinum	Aldosterone	Thymosin
75. Pregnancy	ACTH	Estrogen	Chorion
76. Melatonin	Menstruation	"Third eye"	Semen

Match the term on the left with the proper selection on the right.

Group A

_____ 77. Alpha cells   A. Estrogen

_____ 78. Beta cells   B. Progesterone

_____ 79. Corpus luteum   C. Insulin

_____ 80. Interstitial cells   D. Testosterone

_____ 81. Ovarian follicles   E. Glucagon

Group B

_____ 82. Placenta   A. Melatonin

_____ 83. Pineal   B. ANH

_____ 84. Heart atria   C. Testosterone

_____ 85. Testes   D. Thymosin

_____ 86. Thymus   E. Chorionic gonadotropins

▶ *If you have had difficulty with this section, review pages 207-209.*

APPLYING WHAT YOU KNOW

87. Mrs. Fortner made a routine visit to her physician last week. When the laboratory results came back, the report indicated a high level of chorionic gonadotropin in her urine. What did this mean to Mrs. Fortner?

88. Mrs. Wilcox noticed that her daughter was beginning to take on the secondary sex characteristics of a male. The pediatrician diagnosed the condition as a tumor of an endocrine gland. Where specifically was the tumor located?

Can you find the 16 terms from the chapter in the box of letters? Words may be spelled top to bottom, bottom to top, right to left, left to right, or diagonally.

```
S  S  I  S  E  R  U  I  D  M  E  S  I  T  W
N  D  N  X  E  B  A  M  E  D  E  X  Y  M  I
I  I  G  O  N  R  S  G  X  T  I  I  Y  V  B
D  O  M  S  I  N  I  T  E  R  C  C  U  Q  D
N  C  X  S  R  T  N  B  O  T  V  M  Y  O  M
A  I  M  E  C  L  A  C  R  E  P  Y  H  I  X
L  T  Y  R  O  I  E  Z  Q  R  T  J  V  K  F
G  R  E  T  D  P  S  N  I  L  D  J  M  X  N
A  O  O  S  N  I  D  K  I  N  K  S  P  P  O
T  C  P  R  E  T  I  O  G  R  I  F  M  X  G
S  L  M  H  Y  P  O  G  L  Y  C  E  M  I  A
O  A  J  L  H  O  R  M  O  N  E  O  T  G  C
R  C  E  L  T  S  E  L  C  N  P  N  X  U  U
P  I  O  S  W  R  T  X  C  G  U  L  O  E  L
G  G  V  Y  H  M  S  H  Y  K  A  K  N  Q  G
```

Corticoids	Glucagon	Myxedema
Cretinism	Goiter	Prostaglandins
Diabetes	Hormone	Steroids
Diuresis	Hypercalcemia	Stress
Endocrine	Hypoglycemia	
Exocrine	Luteinization	

DID YOU KNOW?

The total daily output of the pituitary gland is less than 1/1,000,000 of a gram, yet this small amount is responsible for stimulating the majority of all endocrine functions.

THE ENDOCRINE SYSTEM

ACROSS

1. Secreted by cells in the walls of the heart's atria
4. Adrenal medulla
6. Estrogens
8. Converts amino acids to glucose
9. Melanin
11. Labor

DOWN

2. Hypersecretion of insulin
3. Antagonist to diuresis
5. Increases calcium concentration
7. Hyposecretion of Islands of Langerhans (one word)
8. Hyposecretion of thyroid
10. Adrenal cortex

ENDOCRINE GLANDS

1. _____

2. _____

3. _____

4. _____

5. _____

6. _____

7. _____

8. _____

9. _____

10   **Blood**

Blood, the river of life, is the body's primary means of transportation. Although it is the respiratory system that provides oxygen for the body, the digestive system that provides nutrients, and the urinary system that eliminates wastes, none of these functions could be provided for the individual cells without the blood. In less than one minute, a drop of blood will complete a trip through the entire body, distributing nutrients and collecting the wastes of metabolism.

Blood is divided into plasma, which is the liquid portion of blood, and the formed elements, which are the blood cells. There are three types of blood cells: red blood cells, white blood cells, and platelets. Together these cells and plasma provide a means of transportation that delivers the body's daily necessities.

Although the red blood cells in all of us are of a similar shape, we have different blood types. Blood types are identified by the presence of certain antigens in the red blood cells. Every person's blood belongs to one of four main blood groups: Type A, B, AB, or O. Any one of the four groups or "types" may or may not have the Rh factor present in the red blood cells. If an individual has a specific antigen called the Rh factor present in his or her blood, the blood is Rh positive. If this factor is missing, the blood is Rh negative. Approximately 85% of the population have the Rh factor (Rh positive) and 15% do not have the Rh factor (Rh negative).

Your understanding of this chapter will be necessary to prepare a proper foundation for the circulatory system.

TOPICS FOR REVIEW

Before progressing to Chapter 11, you should have an understanding of the structure and function of blood plasma and cells. Your review should also include a knowledge of blood types and Rh factors.

BLOOD COMPOSITION

Circle the best answer.

1. Which one of the following substances is *not* a part of the plasma?
   A. Hormones
   B. Salts
   C. Nutrients
   D. Wastes
   E. All of the above are part of the plasma

2. The normal volume of blood in an adult is about:
   A. 2-3 pints
   B. 2-3 quarts
   C. 2-3 gallons
   D. 4-6 liters

3. Blood is normally:
   A. Very acidic
   B. Slightly acidic
   C. Neutral
   D. Slightly alkaline

4. Another name for white blood cells is:
   A. Erythrocytes
   B. Leukocytes
   C. Thrombocytes
   D. Platelets

5. Another name for platelets is:
   A. Neutrophils
   B. Eosinophils
   C. Thrombocytes
   D. Erythrocytes

6. Pernicious anemia is caused by:
   A. A lack of vitamin B12
   B. Hemorrhage
   C. Radiation
   D. Bleeding ulcers

7. The laboratory test called hematocrit tells the physician:
   A. The volume of white cells in a blood sample
   B. The volume of red cells in a blood sample
   C. The volume of platelets in a blood sample
   D. The volume of plasma in a blood sample

8. An example of a nongranular leukocyte is a/an:
   A. Platelet
   B. Erythrocyte
   C. Eosinophil
   D. Monocyte

9. An abnormally high white blood cell count is known as:
   A. Leukemia
   B. Leukopenia
   C. Leukocytosis
   D. Anemia

10. A critical component of hemoglobin is:
    A. Potassium
    B. Calcium
    C. Vitamin K
    D. Iron

11. Sickle cell anemia is caused by:
    A. The production of an abnormal type of hemoglobin
    B. The production of excessive neutrophils
    C. The production of excessive platelets
    D. The production of abnormal leukocytes

12. The practice of using blood transfusions to increase oxygen delivery to muscles during athletic events is called:
    A. Blood antigen
    B. Blood doping
    C. Blood agglutination
    D. Blood proofing

13. The term used to describe the condition of a circulating blood clot is:
    A. Thrombosis
    B. Embolism
    C. Hemoglobin
    D. Platelet

14. Which one of the following types of cells is not a granular leukocyte?
    A. Neutrophil
    B. Lymphocyte
    C. Basophil
    D. Eosinophil
15. If a blood cell has no nucleus and is shaped like a biconcave disc, then the cell most likely is a/an:
    A. Platelet
    B. Lymphocyte
    C. Basophil
    D. Eosinophil
    E. Red blood cell
16. Red bone marrow forms all kinds of blood cells *except* some:
    A. Platelets
    B. Lymphocytes
    C. Red blood cells
    D. Neutrophils
17. Myeloid tissue is found in all but which one of the following locations?
    A. Sternum
    B. Ribs
    C. Wrist bones
    D. Hip bones
    E. Cranial bones
18. Lymphatic tissue is found in all but which of the following locations?
    A. Lymph nodes
    B. Thymus
    C. Spleen
    D. All of the above contain lymphatic tissue
19. The "buffy coat" layer in a hematocrit tube contains:
    A. Red blood cells and platelets
    B. Plasma only
    C. Platelets only
    D. White blood cells and platelets
    E. None of the above is correct
20. The hematocrit value for red blood cells is _____%.
    A. 75
    B. 60
    C. 50
    D. 45
    E. 35
21. An unusually low white blood cell count would be termed:
    A. Leukemia
    B. Leukopenia
    C. Leukocytosis
    D. Anemia
    E. None of the above is correct
22. Most of the oxygen transported in the blood is carried by:
    A. Platelets
    B. Plasma
    C. Basophils
    D. Red blood cells
    E. None of the above is correct
23. The most numerous of the phagocytes are the _____.
    A. Lymphocytes
    B. Neutrophils
    C. Basophils
    D. Eosinophils
    E. Monocytes
24. Which one of the following types of cells is not phagocytic?
    A. Neutrophils
    B. Eosinophils
    C. Lymphocytes
    D. Monocytes
    E. All of the above are phagocytic cells
25. Which of the following cell types functions in the immune process?
    A. Neutrophils
    B. Lymphocytes
    C. Monocytes
    D. Basophils
    E. Reticuloendothelial cells

26. The organ that manufactures prothrombin is the:
   A. Liver
   B. Pancreas
   C. Thymus
   D. Kidney
   E. Spleen
27. Which one of the following vitamins acts to accelerate blood clotting?
   A. A
   B. B
   C. C
   D. D
   E. K

▶ *If you have had difficulty with this section, review pages 215-222.*

BLOOD TYPES
RH FACTOR

Fill in the blank areas.

28.

Blood Type	Antigen Present in RBC	Antibody Present in Plasma
A	_____	Anti-B
B	B	_____
AB	_____	None
O	None	_____

Fill in the blanks

29. An _____ is a substance that can stimulate the body to make antibodies.
30. An _____ is a substance made by the body in response to stimulation by an antigen.
31. Many antibodies react with their antigens to clump or _____ them.
32. If a baby is born to an Rh negative mother and Rh positive father, it may develop the disease
   _____ _____
33. The term Rh is used because the antigen was first discovered in the blood of a
   _____ _____
34. _____ stops an Rh negative mother from forming anti-Rh antibodies and thus prevents the possibility of harm to the next Rh positive baby.
35. Blood type _____ has been called the universal recipient.

▶ *If you have had difficulty with this section, review pages 222-224.*

APPLYING WHAT YOU KNOW

36. Mrs. Payne's blood type is O positive. Her husband's type is O negative. Her newborn baby's blood type is O negative. Is there any need for concern with this combination?
37. After Mrs. Freund's baby was born, the doctor applied a gauze dressing for a short time on the umbilical cord. He also gave the baby a dose of vitamin K. Why did the doctor perform these two procedures?

38. WORD FIND

Can you find 24 terms from this chapter in the box of letters? Words may be spelled top to bottom, bottom to top, right to left, left to right, or diagonally.

```
H  S  H  K  L  S  U  L  O  B  M  E  A  E  S
K  E  E  V  M  Z  H  E  P  A  R  I  N  D  N
D  T  M  F  A  C  T  O  R  Y  E  T  I  Q  P
O  Y  A  O  H  N  B  A  T  Q  Y  A  R  A  R
N  C  T  J  G  W  E  H  N  P  N  L  B  U  D
O  O  O  S  Q  L  R  M  E  T  I  S  I  P  M
R  K  C  E  M  O  O  N  I  H  I  Q  F  W  W
H  U  R  T  C  N  I  B  P  A  S  G  Z  T  G
E  E  I  Y  O  B  O  O  I  V  E  W  E  T  X
S  L  T  C  M  D  S  E  C  N  R  T  U  N  R
U  E  Y  O  Y  A  I  M  E  K  U  E  L  Y  S
S  T  R  G  B  S  T  H  R  O  M  B  U  S  Q
E  H  Z  A  M  S  A  L  P  N  D  Z  P  O  N
T  E  N  H  F  P  D  A  P  M  E  E  I  B  W
E  W  B  P  H  K  B  O  N  C  K  W  X  V  J
```

AIDS	Factor	Phagocytes
Anemia	Fibrin	Plasma
Antibody	Hematocrit	Recipient
Antigen	Hemoglobin	Rhesus
Basophil	Heparin	Serum
Donor	Leukemia	Thrombin
Embolus	Leukocytes	Thrombus
Erythrocytes	Monocyte	Type

DID YOU KNOW?

Blood products are good for approximately 21 days, whereas fresh frozen plasma is good for at least 6 months.

BLOOD

ACROSS

1. Abnormally high WBC count
4. Final stage of clotting process
6. Oxygen carrying mechanism of blood
9. To engulf and digest microbes
10. Stationary blood clot
12. RBC
13. Circulating blood clot
14. Liquid portion of blood

DOWN

2. Type O (two words)
3. Substances that stimulate the body to make antibodies
5. Type of leukocyte
7. Platelets
8. Prevents clotting of blood
11. Inability of the blood to carry sufficient oxygen

HUMAN BLOOD CELLS

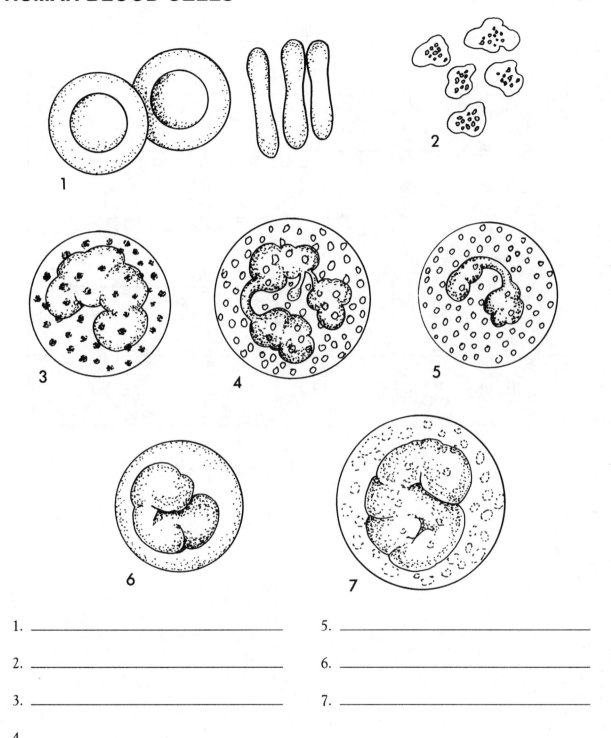

1. _____

2. _____

3. _____

4. _____

5. _____

6. _____

7. _____

BLOOD TYPING

Using the key below, draw the appropriate reaction with the donor's blood in the circles.

Recipient's blood		Reactions with donor's blood			
RBC antigens	Plasma antibodies	Donor type O	Donor type A	Donor type B	Donor type AB
None (Type O)	Anti-A Anti-B	◯	◯	◯	◯
A (Type A)	Anti-B	◯	◯	◯	◯
B (Type B)	Anti-A	◯	◯	◯	◯
AB (Type AB)	(none)	◯	◯	◯	◯

 Normal blood    Agglutinated blood

11   **The Circulatory System**

The heart is actually two pumps, one to move blood to the lungs, the other to push it out into the body. These two functions seem rather elementary by comparison to the complex and numerous functions performed by most of the other body organs, and yet, if this pump stops, within a few short minutes all life ceases.

The heart is divided into two upper compartments called atria or receiving chambers and two lower compartments or discharging chambers called ventricles. By age 45, approximately 300,000 tons of blood will have passed through these chambers to be circulated to the blood vessels. These vessels, called arteries, veins, and capillaries serve different functions. Arteries carry blood from the heart, veins carry blood to the heart, and capillaries are exchange vessels or connecting links between the arteries and veins. This closed system of circulation provides distribution of blood to the whole body (systemic circulation) and to specific regions, such as pulmonary circulation or hepatic portal circulation.

Blood pressure is the force of blood in the vessels. This force is highest in arteries and lowest in veins. Normal blood pressure varies among individuals and depends on the volume of blood in the arteries. The larger the volume of blood in the arteries, the more pressure is exerted on the walls of the arteries, and the higher the arterial pressure. Conversely, the less blood in the arteries, the lower the blood pressure.

A functional cardiovascular system is vital for survival because without circulation, tissues would lack a supply of oxygen and nutrients. Waste products would begin to accumulate and could become toxic. Your review of this system will provide you with an understanding of the complex transportation mechanism of the body necessary for survival.

TOPICS FOR REVIEW

Before progressing to Chapter 12, you should have an understanding of the structure and function of the heart and blood vessels. Your review should include a study of systemic, pulmonary, hepatic portal, and fetal circulations, and should conclude with a thorough understanding of blood pressure and pulse.

HEART

Fill in the blanks.

1. Rhythmic compression of the heart combined with effective artificial respiration is known as _____

2. The _____ _____ divides the heart into right and left sides between the atria.

3. The _____ are the two upper chambers of the heart.

4. The _____ are the two lower chambers of the heart.

5. The cardiac muscle tissue is referred to as the _____.

6. Inflammation of the heart lining is _____.

7. The two AV valves are _____ and _____.

8. _____ _____ involves movement of blood from the right ventricle to the lungs.

9. An occlusion of a coronary artery is known as _____.

10. _____ _____ occurs when heart muscle cells are deprived of oxygen and become damaged or die.

11. The pacemaker of the heart is the _____ node.

12. A normal ECG tracing has three characteristic waves. They are _____, _____, and _____ waves.

13. _____ begins just before the relaxation phase of cardiac muscle activity noted on an ECG.

Select the best answer.

a. Pericardium
b. Severe chest pain
c. Thrombosis
d. Pulmonary
e. Heart block
f. Ventricles
g. Systemic

h. Coronary arteries
i. Systole
j. Depolarization
k. Atria
l. Apex
m. Epicardium

_____ 14. Covering of heart
_____ 15. Receiving chambers
_____ 16. Circulation from left ventricle throughout body
_____ 17. Blood clot
_____ 18. Semilunar valve
_____ 19. Discharging chambers
_____ 20. Supplies oxygen to heart muscle
_____ 21. Angina pectoris
_____ 22. Slow heart rate caused by blocked impulses
_____ 23. Contraction of the heart
_____ 24. Electrical activity associated with ECG
_____ 25. Blunt-pointed lower edge of heart
_____ 26. Visceral pericardium

▶ *If you have had difficulty with this section, review pages 229-238.*

BLOOD VESSELS CIRCULATION

Match the term on the left with the proper selection on the right.

_____ 27. Arteries

_____ 28. Veins

_____ 29. Capillaries

_____ 30. Tunica adventitia

_____ 31. Precapillary sphincters

_____ 32. Superior vena cava

_____ 33. Aorta

A. Smooth muscle cells that guard entrance to capillaries
B. Carry blood to the heart
C. Carry blood into venules
D. Carry blood away from the heart
E. Largest vein
F. Largest artery
G. Outermost layer of arteries and veins

Circle the best answer.

34. The aorta carries blood out of the:
    A. Right atrium
    B. Left atrium
    C. Right ventricle
    D. Left ventricle
    E. None of the above is correct

35. The superior vena cava returns blood to the:
    A. Left atrium
    B. Left ventricle
    C. Right atrium
    D. Right ventricle
    E. None of the above is correct

36. Which one of the following vessels has its wall made up entirely of endothelial cells?
    A. Vein
    B. Capillary
    C. Artery
    D. Venule
    E. Arteriole

37. The _____ is made up of smooth muscle.
    A. Tunica media
    B. Tunica adventitia
    C. Tunica intima
    D. Endothelium
    E. Myocardium

38. The _____ function as exchange vessels.
    A. Venules
    B. Capillaries
    C. Arteries
    D. Arterioles
    E. Veins

39. Blood returns from the lungs during pulmonary circulation via the:
    A. Pulmonary artery
    B. Pulmonary veins
    C. Aorta
    D. Inferior vena cava

40. The hepatic portal circulation serves the body by:
    A. Removing excess glucose and storing it in the liver as glycogen
    B. Detoxifying blood
    C. Removing various poisonous substances present in blood
    D. All of the above

41. The structure used to bypass the liver in fetal circulation is the:
   A. Foramen ovale   C. Ductus arteriosus
   B. Ductus venosus   D. Umbilical vein
42. The foramen ovale serves the fetal circulation by:
   A. Connecting the aorta and the pulmonary artery
   B. Shunting blood from the right atrium directly into the left atrium
   C. Bypassing the liver
   D. Bypassing the lungs
43. The structure used to connect the aorta and pulmonary artery in fetal circulation is the:
   A. Ductus arteriosus   C. Aorta
   B. Ductus venosus   D. Foramen ovale
44. Which of the following is *not* an artery?
   A. Femoral   C. Coronary
   B. Popliteal   D. Inferior vena cava
45. Which of the following has valves to assist the blood flow:
   A. Veins   C. Capillaries
   B. Arteries   D. Arterioles

▶ *If you have had difficulty with this section, review pages 238-248.*

BLOOD PRESSURE PULSE

Mark T if the answer is true. If the answer is false, circle the wrong word(s) and correct the statement by inserting the proper word(s) in the answer blank.

_____ 46. Blood pressure is highest in the veins and lowest in the arteries.
_____ 47. The difference between two blood pressures is referred to as blood pressure deficit.
_____ 48. If the blood pressure in the arteries were to decrease so that it became equal to the average pressure in the arterioles, circulation would increase.
_____ 49. A stroke is often the result of low blood pressure.
_____ 50. Massive hemorrhage increases blood pressure.
_____ 51. Blood pressure is the volume of blood in the vessels.
_____ 52. Both the strength and the rate of heartbeat affect cardiac output and blood pressure.
_____ 53. The diameter of the arterioles helps to determine how much blood drains out of arteries into arterioles.
_____ 54. A stronger heartbeat tends to decrease blood pressure and a weaker heartbeat tends to increase it.
_____ 55. The systolic pressure is the pressure while the ventricles relax.
_____ 56. The diastolic pressure is the pressure while the ventricles contract.
_____ 57. The pulse is a vein expanding and then recoiling.
_____ 58. The radial artery is located at the wrist.

_____ 59. The common carotid artery is located in the neck along the front edge of the sternocleidomastoid muscle.

_____ 60. The artery located at the bend of the elbow and used for locating the pulse is the dorsalis pedis.

▷ *If you have had difficulty with this section, review pages 248-252.*

Unscramble the words.

61. STMESYCI
⬜⭕⭕⬜⬜⬜⬜

62. NULVEE
⭕⬜⬜⬜⭕

63. RYTREA
⬜⬜⬜⭕⬜⬜

64. USLEP
⬜⬜⭕⬜⬜

Take the circled letters, unscramble them, and fill in the statement.
How Noah survived the flood.

65. ⬜⬜⬜⬜⬜⬜⬜

APPLYING WHAT YOU KNOW

66. Elsa was experiencing angina pectoris. Her doctor suggested a surgical procedure that would require the removal of a vein from another region of her body. They would then use the vein to bypass a partial blockage in her coronary arteries. What is this procedure called?

67. Mr. Stuckey has heart block. His electrical impulses are being blocked from reaching the ventricles. An electrical device that causes ventricular contractions at a rate necessary to maintain circulation is being considered as possible treatment for his condition. What is this device?

68. Mrs. Heygood was diagnosed with an acute case of endocarditis. What is the real danger of this diagnosis?

69. Mr. Philbrick returned from surgery in stable condition. The nurse noted that each time she took Mr. Philbrick's pulse and blood pressure, the pulse became higher and the blood pressure lower than the last time. What might be the cause?

Can you find 15 terms from this chapter in the box of letters? Words may be spelled top to bottom, bottom to top, right to left, left to right, or diagonally.

```
Y H Y S I S O B M O R H T A R
L A C I L I B M U O A Y N I Y
E E S Y S T E M I C C G D E U
M V E N U L E D D C I X M D I
E Y M U I R T A R N L E C G Y
N O I T A Z I R A L O P E D N
U D L A T R O P C I T A P E H
S I U K M U E T O E S L U P U
R P N F Y C B E D R A L Z P J
D S A H T I T V N L I I B D W
Q U R O I V A Q E O D A K R K
K C R M P G A I G N D D Z Y J
Y I Y R I N E A F Z L Q S P X
S R M I C C Q O A W U N H O K
P T A V H H Z L H I J J X K Z
```

Angina pectoris   ECG   Systemic

Apex   Endocardium   Thrombosis

Atrium   Hepatic portal   Tricuspid

Depolarization   Pulse   Umbilical

Diastolic   Semilunar   Venule

DID YOU KNOW?

- Your heart pumps more than 5 quarts of blood every minute or 2,000 gallons a day.
- Every pound of excess fat contains some 200 miles of additional capillaries to push blood through each minute.

CIRCULATORY SYSTEM

ACROSS

2. Inflammation of the lining of the heart
3. Bicuspid valve (2 words)
5. Inner layer of pericardium
7. Cardiopulmonary resuscitation (abbreviation)
10. Carries blood away from the heart
11. Upper chamber of heart
12. Lower chambers of the heart
13. SA node

DOWN

1. Unique blood circulation through the liver (2 words)
3. Muscular layer of the heart
4. Carries blood to the heart
6. Tiny artery
8. Heart rate
9. Carries blood from arterioles into venules

THE HEART

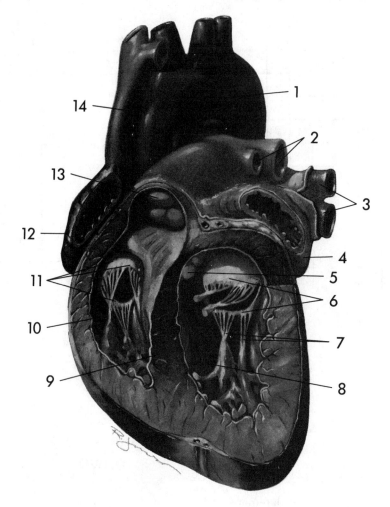

1. _____

2. _____

3. _____

4. _____

5. _____

6. _____

7. _____

8. _____

9. _____

10. _____

11. _____

12. _____

13. _____

14. _____

CONDUCTION SYSTEM OF THE HEART

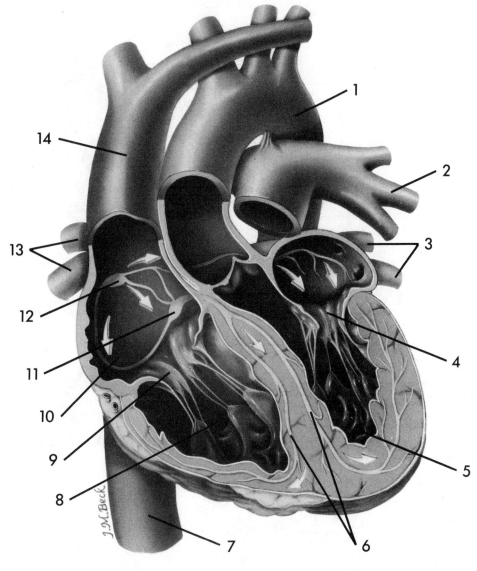

1. _____   8. _____

2. _____   9. _____

3. _____   10. _____

4. _____   11. _____

5. _____   12. _____

6. _____   13. _____

7. _____   14. _____

FETAL CIRCULATION

1. _____
2. _____
3. _____
4. _____
5. _____
6. _____
7. _____
8. _____
9. _____
10. _____
11. _____
12. _____
13. _____
14. _____
15. _____
16. _____
17. _____
18. _____

HEPATIC PORTAL CIRCULATION

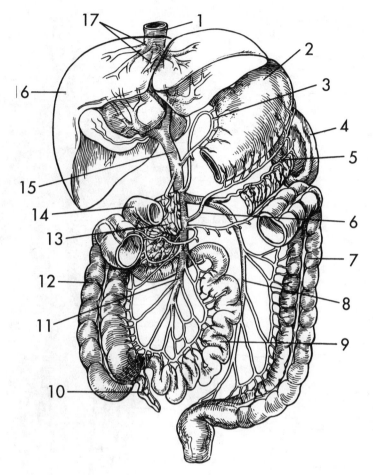

1. _____

2. _____

3. _____

4. _____

5. _____

6. _____

7. _____

8. _____

9. _____

10. _____

11. _____

12. _____

13. _____

14. _____

15. _____

16. _____

17. _____

18. _____

PRINCIPAL ARTERIES OF THE BODY

1. _____
2. _____
3. _____
4. _____
5. _____
6. _____
7. _____
8. _____
9. _____
10. _____
11. _____
12. _____
13. _____
14. _____
15. _____
16. _____
17. _____
18. _____
19. _____

20. _____   24. _____   28. _____
21. _____   25. _____   29. _____
22. _____   26. _____   30. _____
23. _____   27. _____

PRINCIPAL VEINS OF THE BODY

1. _____

2. _____

3. _____

4. _____

5. _____

6. _____

7. _____

8. _____

9. _____

10. _____

11. _____

12. _____

13. _____

14. _____

15. _____

16. _____

17. _____

18. _____

19. _____

20. _____   25. _____   30. _____

21. _____   26. _____   31. _____

22. _____   27. _____   32. _____

23. _____   28. _____   33. _____

24. _____   29. _____   34. _____

NORMAL ECG DEFLECTIONS

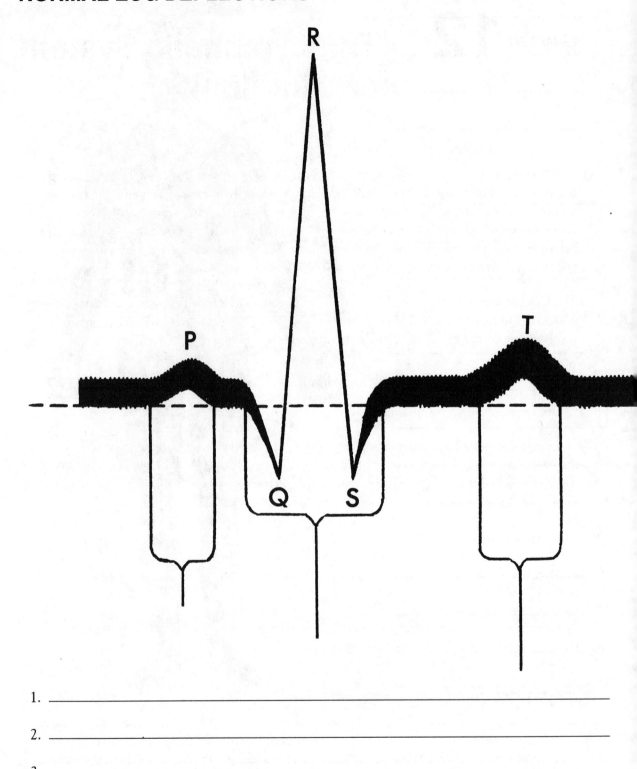

1. _____

2. _____

3. _____

12 # The Lymphatic System and Immunity

The lymphatic system is a system similar to the circulatory system. Lymph, like blood, flows through an elaborate route of vessels. In addition to lymphatic vessels, the lymphatic system consists of lymph nodes, lymph, and the spleen. Unlike the circulatory system, the lymphatic vessels do not form a closed circuit. Lymph flows only once through the vessels before draining into the general blood circulation. This system is a filtering mechanism for microorganisms and serves as a protective device against foreign invaders, such as cancer.

The immune system is the armed forces division of the body. Ready to attack at a moment's notice, the immune system defends us against the major enemies of the body: microorganisms, foreign transplanted tissue cells, and our own cells that have turned malignant.

The most numerous cells of the immune system are the lymphocytes. These cells circulate in the body's fluids seeking invading organisms and destroying them with powerful lymphotoxins, lymphokines, or antibodies.

Phagocytes, another large group of immune system cells, assist with the destruction of foreign invaders by a process known as phagocytosis. Neutrophils, monocytes, and connective tissue cells called *macrophages* use this process to surround unwanted microorganisms, ingest and digest them, rendering them harmless to the body.

Another weapon that the immune system possesses is complement. Normally a group of inactive enzymes present in the blood, complement can be activated to kill invading cells by drilling holes in their cytoplasmic membranes, allowing fluid to enter the cell until it bursts.

Your review of this chapter will give you an understanding of how the body defends itself from the daily invasion of destructive substances.

TOPICS FOR REVIEW

Before progressing to Chapter 13, you should familiarize yourself with the functions of the lymphatic system, the immune system, and the major structures that make up these systems. Your review should include knowledge of lymphatic vessels, lymph nodes, lymph, antibodies, complement, and the development of B and T cells. Your study should conclude with an understanding of the differences in humoral and cell-mediated immunity.

THE LYMPHATIC SYSTEM

Fill in the blanks.

1. _____ is a specialized fluid formed in the tissue spaces that will be transported by way of specialized vessels to eventually reenter the circulatory system.

2. Blood plasma that has filtered out of capillaries into microscopic spaces between cells is called _____.

3. The network of tiny blind-ended tubes distributed in the tissue spaces is called _____.

4. Lymph eventually empties into two terminal vessels called the _____ and the _____.

5. The thoracic duct has an enlarged pouchlike structure called the _____.

6. Lymph is filtered by moving through _____ which are located in clusters along the pathway of lymphatic vessels.

7. Lymph enters the node through four _____ lymph vessels.

8. Lymph exits from the node through a single _____ lymph vessel.

▶ *If you have had difficulty with this section, review pages 259-262.*

THYMUS
TONSILS
SPLEEN

Choose the correct response.

(a) Thymus   (b) Tonsils   (c) Spleen

_____ 9. Palatine, pharyngeal, and lingual are examples
_____ 10. Largest lymphoid organ in the body
_____ 11. Destroys worn out red blood cells
_____ 12. Located in the mediastinum
_____ 13. Serves as a reservoir for blood
_____ 14. T-lymphocytes
_____ 15. Largest at puberty

▶ *If you have had difficulty with this section, review pages 262-263.*

THE IMMUNE SYSTEM

Match the term on the left with the proper selection on the right.

_____ 16. Nonspecific immunity
_____ 17. Inherited immunity
_____ 18. Specific immunity
_____ 19. Acquired immunity
_____ 20. Immunization

A. Inborn immunity
B. Natural immunity
C. General protection
D. Artificial exposure
E. Memory

IMMUNE SYSTEM MOLECULES

Choose the term that applies to each of the following descriptions. Place the letter for the term in the appropriate answer blank.

a. Antibodies	f. Complement fixation
b. Antigen	g. Complement
c. Allergy	h. Humoral immunity
d. Anaphylactic shock	i. Combining site
e. Monoclonal	j. Macrophage

_____ 21. Hypersensitivity of the immune system to harmless antigens
_____ 22. Life-threatening allergic reaction
_____ 23. Type of very specific antibodies produced from a population of identical cells
_____ 24. Protein compounds normally present in the body
_____ 25. Also known as antibody-mediated immunity
_____ 26. Combines with antibody to produce humoral immunity
_____ 27. Antibody
_____ 28. Process of changing molecule shape slightly to expose binding sites
_____ 29. Phagocyte
_____ 30. Inactive proteins in blood

Circle the one that does not belong.

31. Antibody	Allergy	Protein compound	Combining site
32. Antigen	Invading cells	Foreign protein	Complement
33. Monoclonal	Antibodies	Antigen	Specific
34. Allergy	Complement	Anaphylactic shock	Antigen
35. Monoclonal	14	Complement	Proteins

▶ *If you have had difficulty with this section, review pages 263-268.*

IMMUNE SYSTEM CELLS

Circle the best answer.

36. The most numerous cells of the immune system are the:
   A. Monocytes
   B. Eosinophils
   C. Neutrophils
   D. Lymphocytes
   E. Complement

37. The second stage of B cell development changes an immature B cell into a(an):
   A. Plasma cell
   B. Stem cell
   C. Antibody
   D. Activated B cell
   E. Immature B cell

38. Which one of the terms listed below occurs last in the immune process?
   A. Plasma cells
   B. Stem cells
   C. Antibodies
   D. Activated B cells
   E. Immature B cells

39. Moderate exercise has been found to:
    A. Decrease white blood cells
    B. Increase white blood cells
    C. Decrease platelets
    D. Decrease red blood cells

40. Which one of the following is part of the cell membrane of B cells?
    A. Complement   D. Epitopes
    B. Antigens   E. None of the above is correct
    C. Antibodies

41. Immature B cells have:
    A. Four types of defense mechanisms on their cell membrane
    B. Several kinds of defense mechanisms on their cell membrane
    C. One specific kind of defense mechanism on their cell membrane
    D. No defense mechanisms on their cell membrane

42. Development of an immature B cell depends on the B cell coming in contact with:
    A. Complement   D. Lymphokines
    B. Antibodies   E. Antigens
    C. Lymphotoxins

43. The kind of cell that produces large numbers of antibodies is the:
    A. B cell   D. Memory cell
    B. Stem cell   E. Plasma cell
    C. T cell

44. Just one of these short-lived cells that make antibodies can produce _____
    of them per second.
    A. 20   C. 2000
    B. 200   D. 20,000

45. Which of the following statements is *not* true of memory cells?
    A. They produce large numbers of antibodies
    B. They are found in lymph nodes
    C. They develop into plasma cells
    D. They can react with antigens
    E. All of the above are true of memory cells

46. T cell development begins in the:
    A. Lymph nodes   D. Spleen
    B. Liver   E. Thymus
    C. Pancreas

47. Human Immunodeficiency Virus (HIV) has its most obvious effects in:
    A. B cells
    B. Stem cells
    C. Plasma cells
    D. T cells

48. Interferon:
    A. Is produced by T cells within hours after infection by a virus
    B. Decreases the severity of many virus-related diseases
    C. Shows promise as an anticancer agent
    D. Has been shown to be effective in treating breast cancer
    E. All of the above

49. B cells function indirectly to produce:
    A. Humoral immunity   C. Lymphotoxins
    B. Cell-mediated immunity D. Lymphokines
50. T cells function to produce:
    A. Humoral immunity   C. Antibodies
    B. Cell-mediated immunity D. Memory cells

Fill in the blanks.

51. The first stage of development for B cells is called the _____ _____.
52. The second stage of B cell development changes an immature B cell into a/an _____ _____.
53. _____ _____ secrete copious amounts of antibody into the blood—nearly 2000 antibody molecules for every second they live.
54. T cells are lymphocytes that have undergone their first stage of development in the

    _____ _____.
55. _____ blocks HIV's ability to reproduce within infected cells.
56. _____ is a disease caused by a retrovirus that enters the bloodstream and integrates into the DNA of T cell lymphocytes.
57. Like many viruses, such as the common cold, HIV changes rapidly so the development of a _____ may not occur for several years.

▷ *If you have had difficulty with this section, review pages 268-273.*

Unscramble the words.

58. NTCMPEOLEM

59. MTMYIUNI

60. OENCLS

61. FNROERTENI

Take the circled letters, unscramble them, and fill in the statement.

What the student was praying for the night before exams.

62.

APPLYING WHAT YOU KNOW

63. Two-year old baby Metcalfe was exposed to chickenpox. He had been a particularly sickly child and so the doctor decided to give him a dose of Interferon. What effect was the physician hoping for in baby Metcalfe's case?

64. Marcia was a bisexual and an intravenous drug user. She was diagnosed in 1985 as having HTLV-III. What is the current name for this virus?

65. Baby Haskins was born without a thymus gland. Immediate plans were made for a transplant to be performed. In the meantime, baby Haskins was placed in strict isolation. For what reason was he placed in isolation?

66. WORD FIND

Can you find 14 terms from the chapter in the box of letters? Words may be spelled top to bottom, bottom to top, right to left, left to right, or diagonally.

```
I N F L A M M A T O R Y F C G
Q N L O Y Z O C C P A O F S X
M W T A A M N J X Q K R N P H
U R L E R M P X B R U I A L C
C M A C R O P H A G E I E D Z
X M N D K F M N O T U C R N M
A E O R E Z E U O C F B Y E B
Y P L E B G G R H T Y T S E D
Y S C R I S P J O E I T M L A
Z M O T J X E T O N A E E P X
T O N S I L S S U M Y H T S Y
W A O C J N I M W K O Z E N D
D W M K W R M O I P S H Y B G
F H W U J I Z F V D Z L Y T X
```

Acquired   Interferon   Proteins
Antigen   Lymph   Spleen
Humoral   Lymphocytes   Thymus
Immunity   Macrophage   Tonsils
Inflammatory   Monoclonal

DID YOU KNOW?

In the United States, the HIV infection rate is increasing four times faster in women than men. Women tend to underestimate their risk.

LYMPH AND IMMUNITY

ACROSS

1. Largest lymphoid organ in the body
3. Connective tissue cells that are phagocytes
4. Protein compounds normally present in the body
5. Remain in reserve then turn into plasma cells when needed (2 words)
9. Synthetically produced to fight certain diseases
10. Lymph exits the node through this lymph vessel
11. Lymph enters the node through these lymph vessels

DOWN

2. Secretes a copious amount of antibodies into the blood (2 words)
6. Inactive proteins
7. Family of identical cells descended from one cell
8. Type of lymphocyte (humoral immunity – 2 words)
11. Immune deficiency disorder
12. Type of lymphocyte (cell-mediated immunity – 2 words)

B CELL DEVELOPMENT

Complete the diagram:

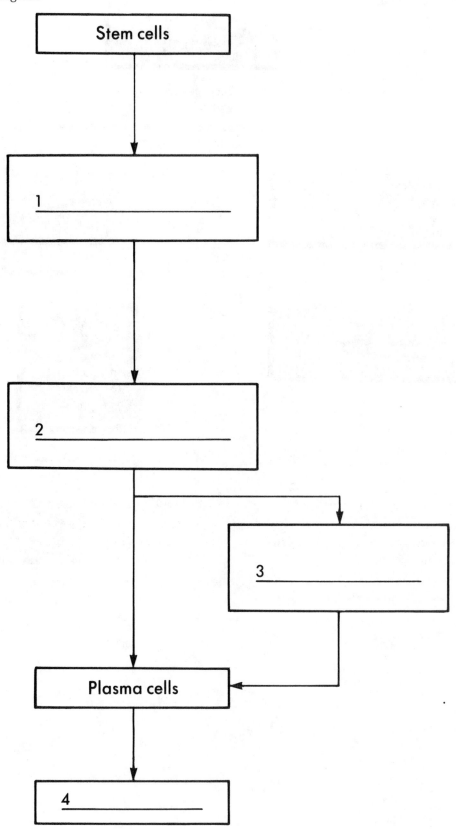

Stem cells

1 _____

2 _____

3 _____

Plasma cells

4 _____

FUNCTION OF SENSITIZED T CELLS

Complete the diagram:

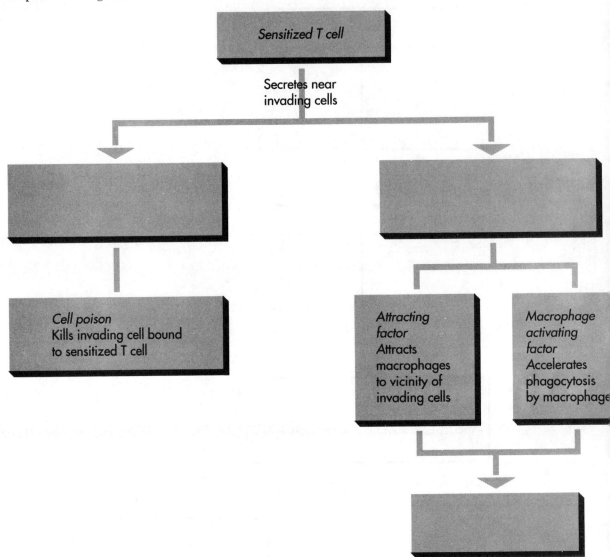

Sensitized T cell

Secretes near
invading cells

Cell poison
Kills invading cell bound
to sensitized T cell

Attracting
factor
Attracts
macrophages
to vicinity of
invading cells

Macrophage
activating
factor
Accelerates
phagocytosis
by macrophage

FUNCTION OF ANTIBODIES

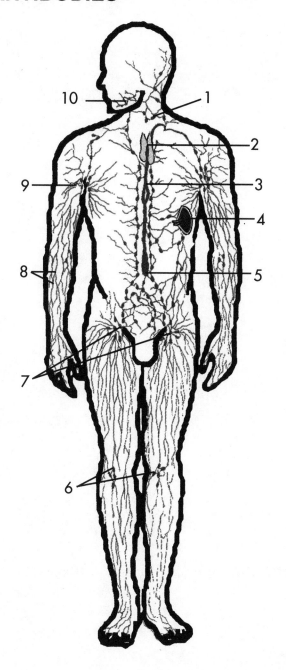

1. _____

2. _____

3. _____

4. _____

5. _____

6. _____

7. _____

8. _____

9. _____

10. _____

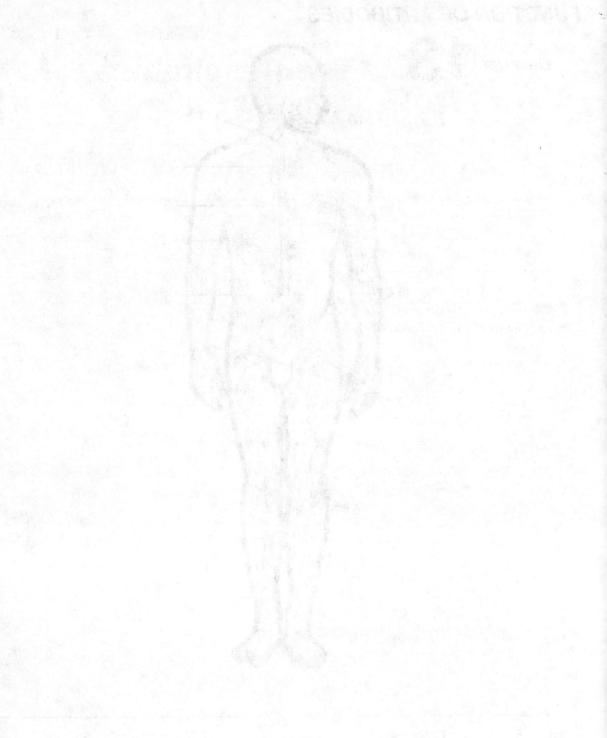

CHAPTER **13**   # The Respiratory System

As you sit reviewing this system, your body needs 16 quarts of air per minute. Walking requires 24 quarts of air, and running requires 50 quarts per minute. The respiratory system provides the air necessary for you to perform your daily activities and eliminates the waste gases from the air that you breathe. Take a deep breath, and think of the air as entering some 250 million tiny air sacs similar in appearance to clusters of grapes. These microscopic air sacs expand to let air in and contract to force it out. These tiny sacs, or alveoli, are the functioning units of the respiratory system. They provide the necessary volume of oxygen and eliminate carbon dioxide 24 hours a day.

Air enters either through the mouth or the nasal cavity. It next passes through the pharynx, past the epiglottis, through the glottis, and the rest of the larynx. It then continues down the trachea, into the bronchi to the bronchioles, and finally through the alveoli. The reverse occurs for expelled air.

The exchange of gases between air in the lungs and in the blood is known as external respiration. The exchange of gases that occurs between the blood and the cells of the body is known as internal respiration. By constantly supplying adequate oxygen and by removing carbon dioxide as it forms, the respiratory system helps to maintain an environment conducive to maximum cell efficiency.

Your review of this system is necessary to provide you with an understanding of this essential homeostatic mechanism.

TOPICS FOR REVIEW

Before progressing to Chapter 14, you should have an understanding of the structure and function of the organs of the respiratory system. Your review should include knowledge of the mechanisms responsible for both internal and external respiration. Your study should conclude with a knowledge of the volumes of air exchanged in pulmonary ventilation and an understanding of how respiration is regulated.

STRUCTURAL PLAN
RESPIRATORY TRACTS
RESPIRATORY MUCOSA

Match the term with the definition.

a. Diffusion
b. Respiratory membrane
c. Alveoli
d. URI
e. Respiration

f. Respiratory mucosa
g. Upper respiratory tract
h. Lower respiratory tract
i. Cilia
j. Air distributor

_____ 1. Function of respiratory system
_____ 2. Pharynx
_____ 3. Passive transport process responsible for actual exchange of gases
_____ 4. Assists with the movement of mucus toward the pharynx
_____ 5. Barrier between the blood in the capillaries and the air in the alveolus
_____ 6. Lines the tubes of the respiratory tree
_____ 7. Terminal air sacs
_____ 8. Trachea
_____ 9. Head cold
_____ 10. Homeostatic mechanism

Fill in the blanks.

The organs of the respiratory system are designed to perform two basic functions. They serve as an: (11) _____ _____ and as a (12) _____ _____. In addition to the above, the respiratory system (13) _____, (14) _____, and (15) _____ the air we breathe. Respiratory organs include the (16) _____, (17) _____, (18) _____, (19) _____, (20) _____, and the (21) _____. The respiratory system ends in millions of tiny, thin-walled sacs called (22) _____. (23) _____ of gases takes place in these sacs. Two aspects of the structure of these sacs assist them in the exchange of gases. First, an extremely thin membrane, the (24) _____ _____ allows for easy exchange, and second, the large number of air sacs makes an enormous (25) _____ area.

▶ *If you have had difficulty with this section, review pages 279-282.*

NOSE
PHARYNX
LARYNX

Circle the word or phrase that does not belong.

26. Nares	Septum	Oropharynx	Conchae
27. Conchae	Frontal	Maxillary	Sphenoidal
28. Oropharynx	Throat	5 inches	Epiglottis
29. Pharyngeal	Adenoids	Uvula	Nasopharynx
30. Middle ear	Tubes	Nasopharynx	Larynx
31. Voice box	Thyroid cartilage	Tonsils	Vocal cords
32. Palatine	Eustachian tube	Tonsils	Oropharynx
33. Pharynx	Epiglottis	Adam's apple	Voice box

Choose the correct response.

    (a) Nose   (b) Pharynx   (c) Larynx

_____ 34. Warms and humidifies air   _____ 38. Septum

_____ 35. Air and food pass through here   _____ 39. Tonsils

_____ 36. Sinuses   _____ 40. Middle ear infections

_____ 37. Conchae   _____ 41. Epiglottis

▶ *If you have had difficulty with this section, review pages 283-284.*

TRACHEA
BRONCHI, BRONCHIOLES, AND ALVEOLI
LUNGS AND PLEURA

Fill in the blanks.

42. The windpipe is more properly referred to as the _____.

43. _____ keep the framework of the trachea almost noncollapsible.

44. A lifesaving technique designed to free the trachea of ingested food or foreign objects is the

_____.

45. The first branch or division of the trachea leading to the lungs is the _____ _____.

46. Each alveolar duct ends in several _____ _____.

47. The narrow part of each lung, up under the collarbone, is its _____.

48. The _____ covers the outer surface of the lungs and lines the inner surface of the rib cage.

49. Inflammation of the lining of the thoracic cavity is _____.

50. The presence of air in the pleural space on one side of the chest is a _____.

▶ *If you have had difficulty with this section, review pages 284-290.*

RESPIRATION

Mark T if the answer is true. If the answer is false, circle the incorrect word(s) and correct the statement.

_____ 51. Diffusion is the process that moves air into and out of the lungs.

_____ 52. For inspiration to take place, the diaphragm and other respiratory muscles relax.

_____ 53. Diffusion is a passive process that results in movement up a concentration gradient.

_____ 54. The exchange of gases that occurs between blood in tissue capillaries and the body cells is external respiration.

_____ 55. Many pulmonary volumes can be measured as a person breathes into a spirometer.

_____ 56. Ordinarily we take about 2 pints of air into our lungs.

_____ 57. The amount of air normally breathed in and out with each breath is called tidal volume.

_____ 58. The largest amount of air that one can breathe out in one expiration is called residual volume.

_____ 59. The inspiratory reserve volume is the amount of air that can be forcibly inhaled after a normal inspiration.

▷ *If you have had difficulty with this section, review pages 290-293.*

Circle the best answer.

60. The term that means the same thing as breathing is:
    A. Gas exchange
    B. Respiration
    C. Inspiration
    D. Expiration
    E. Pulmonary ventilation

61. Carbaminohemoglobin is formed when _____ binds to hemoglobin.
    A. Oxygen
    B. Amino acids
    C. Carbon dioxide
    D. Nitrogen
    E. None of the above is correct

62. Most of the oxygen transported by the blood is:
    A. Dissolved in white blood cells
    B. Bound to white blood cells
    C. Bound to hemoglobin
    D. Bound to carbaminohemoglobin
    E. None of the above is correct

63. Which of the following would *not* assist inspiration?
    A. Elevation of the ribs
    B. Elevation of the diaphragm
    C. Contraction of the diaphragm
    D. Chest cavity becomes longer from top to bottom

64. A young adult male would have a vital capacity of about _____ ml.
    A. 500
    B. 1200
    C. 3300
    D. 4800
    E. 6200

65. The amount of air that can be forcibly exhaled after expiring the tidal volume is known as the:
    A. Total lung capacity
    B. Vital capacity
    C. Inspiratory reserve volume
    D. Expiratory reserve volume
    E. None of the above is correct
66. Which one of the following is correct?
    A. $VC = TV - IRV + ERV$
    B. $VC = TV + IRV - ERV$
    C. $VC = TV + IRV \times ERV$
    D. $VC = TV + IRV + ERV$
    E. None of the above is correct

▷ *If you have had difficulty with this section, review pages 293-297.*

REGULATION OF RESPIRATION
RECEPTORS INFLUENCING RESPIRATION
TYPES OF BREATHING

Match the term on the left with the proper selection on the right.

_____ 67. Inspiratory center	a. Difficult breathing
_____ 68. Chemoreceptors	b. Located in carotid bodies
_____ 69. Pulmonary stretch receptors	c. Slow and shallow respirations
_____ 70. Dyspnea	d. Normal respiratory rate
_____ 71. Respiratory arrest	e. Located in the medulla
_____ 72. Eupnea	f. Failure to resume breathing following a period of apnea
_____ 73. Hypoventilation	g. Located throughout pulmonary airways and in the alveoli

▷ *If you have had difficulty with this section, review pages 297-299.*

Unscramble the words.

74. SPUELIRY

    ☐ ◯ ☐ ☐ ◯ ☐ ☐

75. CRNBOSITHI

    ☐ ◯ ☐ ◯ ☐ ◯ ☐ ☐ ☐ ☐

76. SESXPTIAI

    ☐ ☐ ☐ ◯ ◯ ☐ ☐ ◯ ☐

77. DDNEAOIS

    ◯ ☐ ◯ ☐ ◯ ☐ ☐ ☐

Take the circled letters, unscramble them, and fill in the statement.

What Mona Lisa was to DaVinci.

78. ☐☐☐☐☐☐☐☐☐☐☐☐☐

APPLYING WHAT YOU KNOW

79. Mr. Gorski is a heavy smoker. Recently he has noticed that when he gets up in the morning, he has a bothersome cough that brings up a large accumulation of mucus. This cough persists for several minutes and then leaves until the next morning. What is an explanation for this problem?

80. Kim was 5 years old and was a mouth breather. She had repeated episodes of tonsillitis and the pediatrician suggested removal of her tonsils and adenoids. He further suggested that the surgery would probably cure her mouth breathing problem. Why is this a possibility?

81. WORD FIND

Can you find 14 terms from this chapter in the box of letters? Words may be spelled top to bottom, bottom to top, right to left, left to right, or diagonally.

```
N  K  S  A  Q  B  I  L  V  A  D  T  I  X  D
O  X  O  O  B  F  I  F  G  I  M  N  R  G  Y
I  G  N  X  W  D  E  E  F  L  B  A  U  T  S
T  B  K  Y  N  H  E  F  O  I  U  T  I  R  P
A  T  M  H  E  O  U  T  R  C  C  C  P  V  N
L  L  A  E  P  S  I  R  I  Z  A  A  N  F  E
I  P  T  M  I  H  G  T  I  P  R  F  P  C  A
T  U  B  O  G  Q  E  V  A  Q  O  R  V  N  W
N  L  N  G  L  R  J  C  C  R  T  U  P  D  C
E  M  U  L  O  V  L  A  U  D  I  S  E  R  E
V  O  V  O  T  A  N  G  J  E  D  P  Z  U  U
O  N  K  B  T  C  N  U  U  E  B  O  S  J  S
P  A  L  I  I  K  E  U  C  N  O  Z  K  N  A
Y  R  V  N  S  D  I  O  N  E  D  A  M  F  I
H  Y  O  M  L  O  A  Z  D  T  Y  M  N  L  K
```

Adenoids	Epiglottis	Residual volume
Carotid body	Hypoventilation	Surfactant
Cilia	Inspiration	URI
Diffusion	Oxyhemoglobin	Vital capacity
Dyspnea	Pulmonary	

DID YOU KNOW?

If the alveoli in our lungs were flattened out, they would cover half of a tennis court.

RESPIRATORY SYSTEM

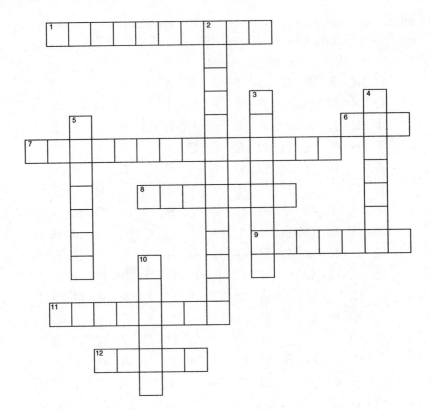

ACROSS

1. Device used to measure the amount of air exchanged in breathing
6. Expiratory reserve volume (abbreviation)
7. Sphenoidal (two words)
8. Terminal air sacs
9. Shelf-like structures that protrude into the nasal cavity
11. Inflammation of pleura
12. Respirations stop

DOWN

2. Surgical procedure to remove tonsils
3. Doctor who developed lifesaving technique
4. Windpipe
5. Trachea branches into right and left structures
10. Voice box

SAGITTAL VIEW OF FACE AND NECK

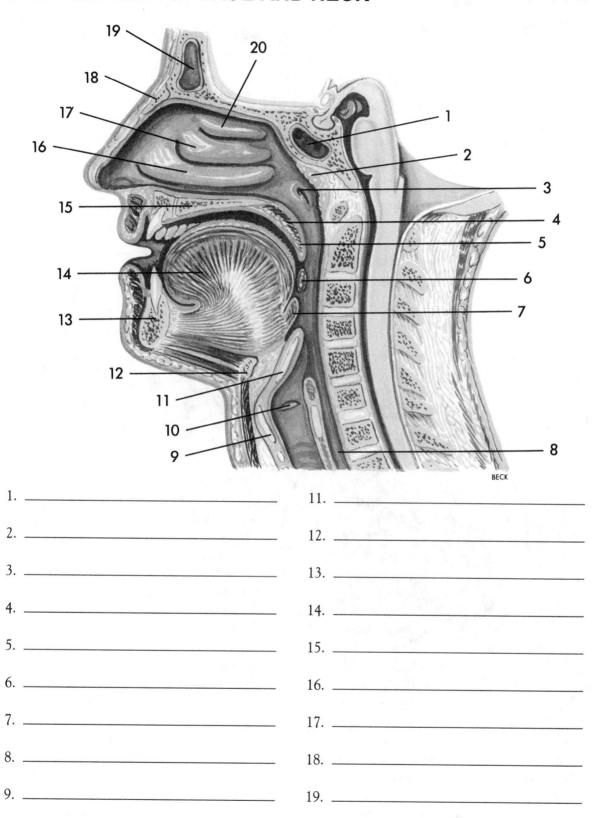

1. _____

2. _____

3. _____

4. _____

5. _____

6. _____

7. _____

8. _____

9. _____

10. _____

11. _____

12. _____

13. _____

14. _____

15. _____

16. _____

17. _____

18. _____

19. _____

20. _____

RESPIRATORY ORGANS

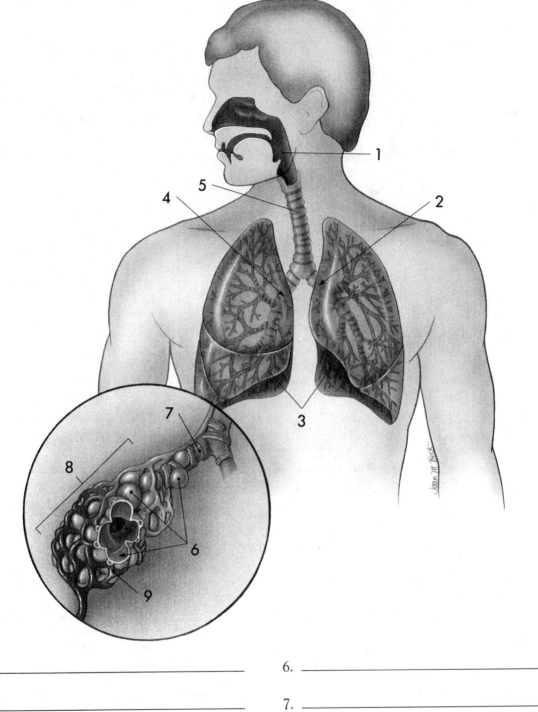

1. _____

2. _____

3. _____

4. _____

5. _____

6. _____

7. _____

8. _____

9. _____

PULMONARY VENTILATION VOLUMES

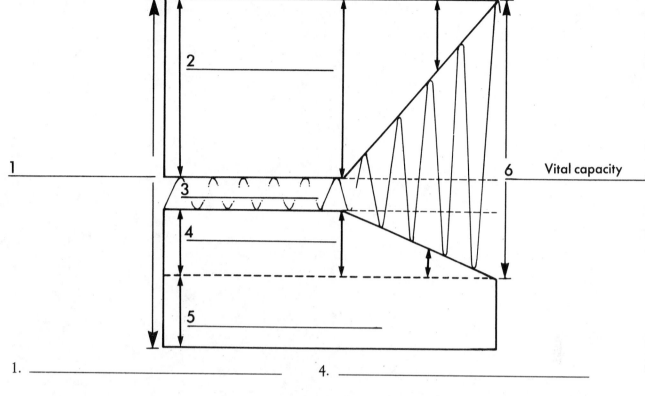

1. _____
2. _____
3. _____
4. _____
5. _____

CHAPTER **14** The Digestive System

Think of the last meal you ate—the different shapes, sizes, tastes, and textures that you so recently enjoyed. Think of those items circulating in your bloodstream in those same original shapes and sizes. Impossible? Of course. And because of this impossibility you will begin to understand and marvel at the close relationship of the digestive system to the circulatory system. It is the digestive system that changes our food, both mechanically and chemically, into a form that is acceptable to the blood and the body.

This change begins the moment you take the very first bite. Digestion starts in the mouth, where food is chewed and mixed with saliva. It then moves down the pharynx and esophagus by peristalsis and enters the stomach. In the stomach it is churned and mixed with gastric juices to become chyme. The chyme goes from the stomach into the duodenum where it is further broken down chemically by intestinal fluids, bile, and pancreatic juice. Those secretions prepare the food for absorption all along the course of the small intestine. Products that are not absorbed pass through the entire length of the small intestine (duodenum, jejunum, ileum). From there they enter into the cecum of the large intestine, the ascending colon, transverse colon, descending colon, sigmoid colon, into the rectum, and out the anus.

Products that are used in the cells undergo absorption. Absorption allows newly processed nutrients to pass through the walls of the digestive tract and into the bloodstream to be distributed to the cells.

Your review of this system will help you understand the mechanical and chemical processes necessary to convert food into energy sources and compounds necessary for survival.

TOPICS FOR REVIEW

Before progressing to Chapter 15, you should review the structure and function of all the organs of digestion. You should have an understanding of the process of digestion, both chemical and mechanical, and of the processes of absorption and metabolism.

WALL OF THE DIGESTIVE SYSTEM

Fill in the blanks.

1. The organs of the digestive system form an irregular-shaped tube called the alimentary canal or _____.

2. The churning of food in the stomach is an example of the _____ breakdown of food.

3. _____ breakdown occurs when digestive enzymes act on food as it passes through the digestive tract.

4. Waste material resulting from the digestive process is known as _____.

5. Foods undergo three kinds of processing in the body: _____, _____, and _____.

6. The serosa of the digestive tube is composed of the _____ _____ in the abdominal cavity.

7. The digestive tract extends from the _____ to the _____.

8. The inside or hollow space within the alimentary canal is called the _____.

9. The inside layer of the digestive tract is the _____.

10. The connective tissue layer that lies beneath the lining of the digestive tract is the _____.

11. The muscularis contracts and moves food through the gastrointestinal tract by a process known as _____.

12. The outermost covering of the digestive tube is the _____.

13. The loops of the digestive tract are anchored to the posterior wall of the abdominal cavity by the _____.

Select the correct response from the two choices given and insert the letter in the answer blank.

(a) Main organ   (b) Accessory organ

_____ 14. Mouth
_____ 15. Parotids
_____ 16. Liver
_____ 17. Stomach
_____ 18. Cecum
_____ 19. Esophagus
_____ 20. Rectum
_____ 21. Pharynx
_____ 22. Appendix
_____ 23. Teeth
_____ 24. Gallbladder
_____ 25. Pancreas

▶ *If you have had difficulty with this section, review pages 305-307.*

MOUTH
TEETH
SALIVARY GLANDS

Circle the best answer.

26. Which one of the following is *not* a part of the roof of the mouth?
    A. Uvula
    B. Palatine bones
    C. Maxillary bones
    D. Soft palate
    E. All of the above are part of the roof of the mouth

27. The largest of the papillae on the surface of the tongue are the:
    A. Filiform
    B. Fungiform
    C. Vallate
    D. Taste buds

28. The first baby tooth, on an average, appears at:
    A. 2 months
    B. 1 year
    C. 3 months
    D. 1 month
    E. 6 months

29. The portion of the tooth that is covered with enamel is the:
    A. Pulp cavity
    B. Neck
    C. Root
    D. Crown
    E. None of the above is correct

30. The wall of the pulp cavity is surrounded by:
    A. Enamel
    B. Dentin
    C. Cementum
    D. Connective tissue
    E. Blood and lymphatic vessels

31. Which of the following teeth is missing from the deciduous arch?
    A. Central incisor
    B. Canine
    C. Second premolar
    D. First molar
    E. Second molar

32. The permanent central incisor erupts between the ages of _____.
    A. 9-13
    B. 5-6
    C. 7-10
    D. 7-8
    E. None of the above is correct

33. The third molar appears between the ages of _____.
    A. 10-14
    B. 5-8
    C. 11-16
    D. 17-24
    E. None of the above is correct

34. Which one of the following will *not* significantly reduce caries?
    A. Reducing the amount of refined sugar in the diet
    B. Fluoride in the water supply
    C. Regular and thorough brushing
    D. Eating a carrot or stick of celery instead of brushing

35. The ducts of the _____ glands open into the floor of the mouth.
    A. Sublingual
    B. Submandibular
    C. Parotid
    D. Carotid

36. The volume of saliva secreted per day is about:
    A. One half pint
    B. One pint
    C. One liter
    D. One gallon

37. Mumps are an infection of the:
    A. Parotid gland
    B. Sublingual gland
    C. Submandibular gland
    D. Tonsils
38. Incisors are used during mastication to:
    A. Cut
    B. Piece
    C. Tear
    D. Grind
39. Another name for the third molar is:
    A. Central incisor
    B. Wisdom tooth
    C. Canine
    D. Lateral incisor
40. After food has been chewed, it is formed into a small, rounded mass called a:
    A. Moat
    B. Chyme
    C. Bolus
    D. Protease

▶ *If you have had difficulty with this section, review pages 307-312.*

PHARYNX
ESOPHAGUS
STOMACH

Fill in the blanks.

The (41) _____ is a tubelike structure that functions as part of both respiratory and digestive systems. It connects the mouth with the (42) _____. The esophagus serves as a passageway for movement of food from the pharynx to the (43) _____. Food enters the stomach by passing through the muscular (44) _____ _____ at the end of the esophagus. Contraction of the stomach mixes the food thoroughly with the gastric juices and breaks it down into a semisolid mixture called (45) _____.

The three divisions of the stomach are the (46) _____, (47) _____, and (48) _____.

Food is held in the stomach by the (49) _____ _____ muscle long enough for partial digestion to occur. After food has been in the stomach for approximately 3 hours, the chyme will enter the (50) _____ _____.

Match the term with the correct definition.

    a. Esophagus
    b. Chyme
    c. Peristalsis
    d. Rugae
    e. Ulcer
    f. Greater curvature
    g. Emesis
    h. Tagamet
    i. Duodenum
    j. Lesser curvature

_____ 51. Stomach folds
_____ 52. Upper right border of stomach
_____ 53. Total emptying of stomach contents back through the cardiac sphincter, up the esophagus, and out of the mouth

_____ 54. 10-inch passageway

_____ 55. Drug used to treat ulcers

_____ 56. Semisolid mixture of stomach contents

_____ 57. Muscle contractions of the digestive system

_____ 58. Open wound in digestive system that is acted on by acidic gastric juice

_____ 59. First part of small intestine

_____ 60. Lower left border of stomach

▶ *If you have had difficulty with this section, review pages 312-314.*

SMALL INTESTINE
LIVER AND GALLBLADDER
PANCREAS

Circle the best answer.

61. Which one is *not* part of the small intestine?
    A. Jejunum
    B. Ileum
    C. Cecum
    D. Duodenum

62. Which one of the following structures does not increase the surface area of the intestine for absorption?
    A. Plicae
    B. Rugae
    C. Villi
    D. Brush border

63. The union of the cystic duct and hepatic duct form the:
    A. Common bile duct
    B. Major duodenal papilla
    C. Minor duodenal papilla
    D. Pancreatic duct

64. Obstruction of the _____ will lead to jaundice.
    A. Hepatic duct
    B. Pancreatic duct
    C. Cystic duct
    D. None of the above

65. Each villus in the intestine contains a lymphatic vessel or _____ that serves to absorb lipid or fat materials from the chyme
    A. Plica
    B. Lacteal
    C. Villa
    D. Microvilli

66. The middle third of the duodenum contains the:
    A. Islets
    B. Fundus
    C. Body
    D. Rugae
    E. Major duodenal papilla

67. Excessive secretion of acid or _____ is an important factor in the formation of ulcers.
    A. Hypoacidity
    B. Emesis
    C. Cholecystokinin
    D. Hyperacidity

68. The liver is an:
    A. Enzyme
    B. Endocrine organ
    C. Endocrine gland
    D. Exocrine gland

69. Fats in chyme stimulate the secretion of the hormone:
    A. Lipase
    B. Cholecystokinin
    C. Protease
    D. Amylase
70. The largest gland in the body is the:
    A. Pituitary
    B. Thyroid
    C. Liver
    D. Thymus

▷ *If you have had difficulty with this section, review pages 314-318.*

LARGE INTESTINE
APPENDIX
PERITONEUM

If the statement is true, mark T next to the answer. If the statement is false, circle the incorrect word(s) and write the correct term in the blank next to the statement.

_____ 71. Bacteria in the large intestine are responsible for the synthesis of vitamin E needed for normal blood clotting.
_____ 72. Villi in the large intestine absorb salts and water.
_____ 73. If waste products pass rapidly through the large intestine, constipation results.
_____ 74. The ileocecal valve opens into the sigmoid colon.
_____ 75. The splenic flexure is the bend between the ascending colon and the transverse colon.
_____ 76. The splenic colon is the S-shaped segment that terminates in the rectum.
_____ 77. The appendix serves no important digestive function in humans.
_____ 78. For patients with suspected appendicitis, a physician will often evaluate the appendix by a digital rectal examination.
_____ 79. The visceral layer of the peritoneum lines the abdominal cavity.
_____ 80. The greater omentum is shaped like a fan and serves to anchor the small intestine to the posterior abdominal wall.

▷ *If you have had difficulty with this section, review pages 318-322.*

DIGESTION
ABSORPTION
METABOLISM

Circle the best answer.
81. Which one of the following substances does *not* contain any enzymes?
    A. Saliva
    B. Bile
    C. Gastric juice
    D. Pancreatic juice
    E. Intestinal juice

82. Which one of the following is a simple sugar?
    A. Maltose
    B. Sucrose
    C. Lactose
    D. Glucose
    E. Starch
83. Cane sugar is the same as:
    A. Maltose
    B. Lactose
    C. Sucrose
    D. Glucose
    E. None of the above is correct
84. Most of the digestion of carbohydrates takes place in the:
    A. Mouth
    B. Stomach
    C. Small intestine
    D. Large intestine
85. Fats are broken down into:
    A. Amino acids
    B. Simple sugars
    C. Fatty acids
    D. Disaccharides

▷ *If you have had difficulty with this section, review pages 322-323.*

CHEMICAL DIGESTION

86. Fill in the blank areas on the chart below.

DIGESTIVE JUICES AND ENZYMES	SUBSTANCE DIGESTED (OR HYDROLYZED)	RESULTING PRODUCT
SALIVA		
1. Amylase	1. _____	1. Maltose
GASTRIC JUICE		
2. Protease (pepsin) plus hydrochloric aicd	2. Proteins	2. _____
PANCREATIC JUICE		
3. Protease (trypsin)	3. Proteins (intact or partially digested)	3. _____
4. Lipase	4. _____	4. Fatty acids, monoglycerides, and glycerol
5. Amylase	5. _____	5. Maltose
INTESTINAL JUICE		
6. Peptidases	6. _____	6. Amino acids
7. _____	7. Sucrose	7. Glucose and fructose
8. Lactase	8. _____	8. Glucose and galactose (simple sugars)
9. Maltase	9. Maltose	9. _____

▷ *If you have had difficulty with this section, review pages 322-323 and Table 14-2.*

APPLYING WHAT YOU KNOW

87. Mr. Amato was a successful businessman, but he worked too hard and was always under great stress. His doctor cautioned him that if he did not alter his style of living he would be subject to hyperacidity. What could be the resulting condition of hyperacidity?

88. Baby Shearer has been regurgitating his bottle feeding at every meal. The milk is curdled, but does not apper to be digested. He has become dehydrated, and so his mother is taking him to the pediatrician. What is a possible diagnosis from your textbook reading?

89. Mr. Attanas has gained a great deal of weight suddenly. He has also noticed that he is sluggish and always tired. What test might his physician order for him and for what reason?

90. WORD FIND

Can you find the 22 terms from the chapter in the box of letters? Words may be spelled top to bottom, bottom to top, right to left, left to right, or diagonally.

```
X   M   E   T   A   B   O   L   I   S   M   X   X   W
R   S   D   P   E   R   I   S   T   A   L   S   I   S
E   V   A   M   N   O   I   T   S   E   G   I   D   R
E   D   E   E   U   O   Q   T   W   Q   O   H   N   Q
Q   Z   H   S   R   N   I   N   T   F   E   C   E   S
D   H   R   E   C   C   I   T   K   C   J   A   P   E
Q   W   R   N   A   T   N   V   P   Y   R   M   P   C
O   C   A   T   N   R   B   A   P   R   H   O   A   I
U   Y   I   E   O   W   T   A   P   A   O   T   W   D
B   O   D   R   L   N   P   B   F   Q   J   S   V   N
N   T   S   Y   F   I   S   L   U   M   E   E   B   U
W   G   J   A   L   U   V   U   N   R   N   W   O   A
Q   S   N   L   X   D   U   O   D   E   N   U   M   J
H   C   A   V   I   T   Y   M   U   C   O   S   A   D
Y   E   A   A   P   H   V   W   S   V   C   Q   J   C
```

Absorption	Emulsify	Mucosa
Appendix	Feces	Pancreas
Cavity	Fundus	Papillae
Crown	Heartburn	Peristalsis
Dentin	Jaundice	Stomach
Diarrhea	Mastication	Uvula
Digestion	Mesentery	
Duodenum	Metabolism	

DID YOU KNOW?

The liver performs over 500 functions and produces over 1000 enzymes to handle the chemical conversions necessary for survival.

DIGESTIVE SYSTEM

ACROSS

5. Digested food moves from intestine to blood
8. Semisolid mixture
9. Inflammation of the appendix
11. Rounded mass of food
13. Stomach folds

DOWN

1. Yellowish skin discoloration
2. Process of chewing
3. Fluid stools
4. Movement of food through digestive tract
6. Vomitus
7. Waste product of digestion
10. Intestinal folds
12. Open wound in digestive area acted on by acid juices
14. Prevents food from entering nasal cavities

DIGESTIVE ORGANS

1. _____

2. _____

3. _____

4. _____

5. _____

6. _____

7. _____

8. _____

9. _____

10. _____

11. _____

12. _____

13. _____

14. _____

15. _____

16. _____

17. _____

18. _____

19. _____

20. _____

21. _____

22. _____

23. _____

24. _____

25. _____

26. _____

27. _____

28. _____

29. _____

30. _____

TOOTH

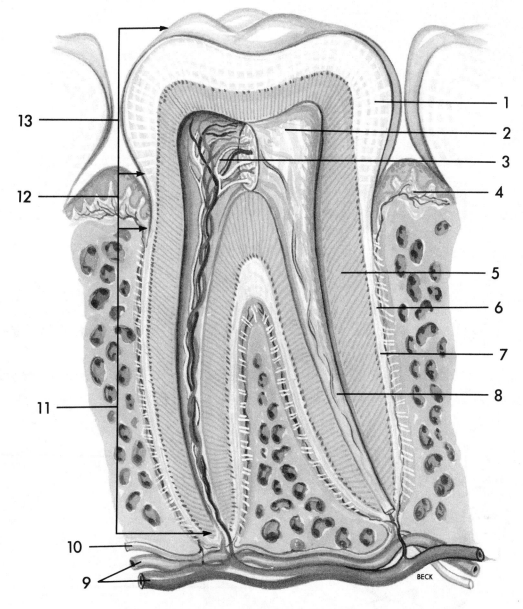

1. _____

2. _____

3. _____

4. _____

5. _____

6. _____

7. _____

8. _____

9. _____

10. _____

11. _____

12. _____

13. _____

THE SALIVARY GLANDS

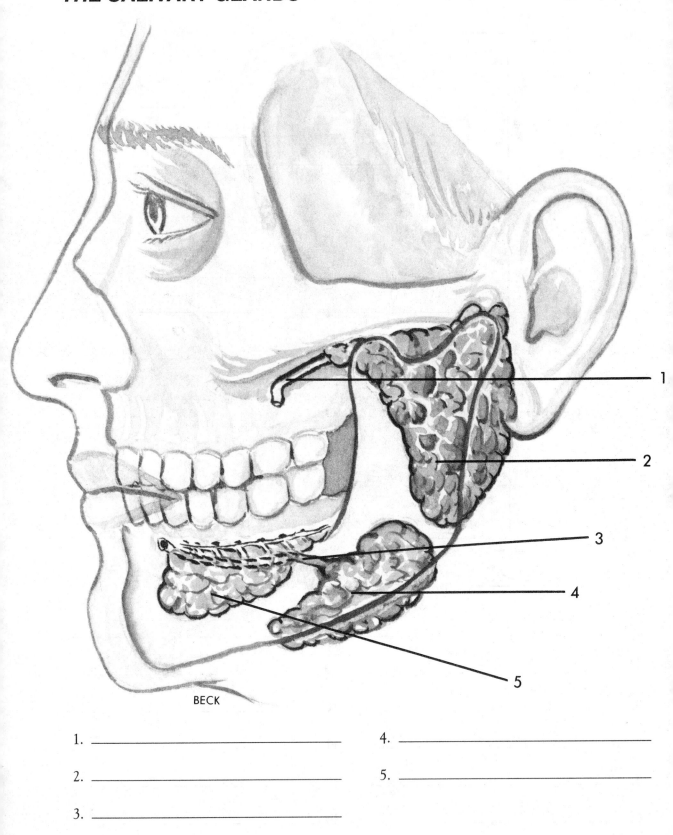

BECK

1. _____ 4. _____

2. _____ 5. _____

3. _____

STOMACH

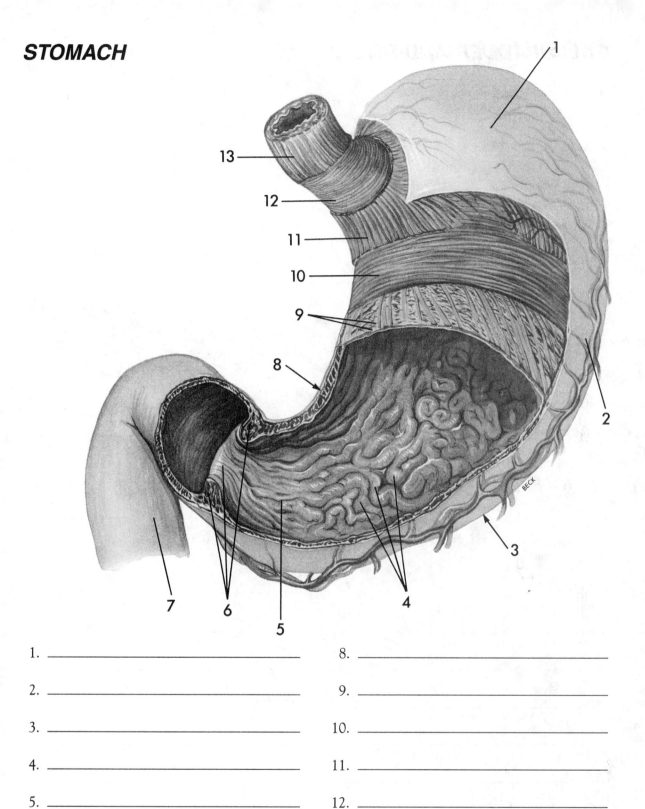

1. _____

2. _____

3. _____

4. _____

5. _____

6. _____

7. _____

8. _____

9. _____

10. _____

11. _____

12. _____

13. _____

GALLBLADDER AND BILE DUCTS

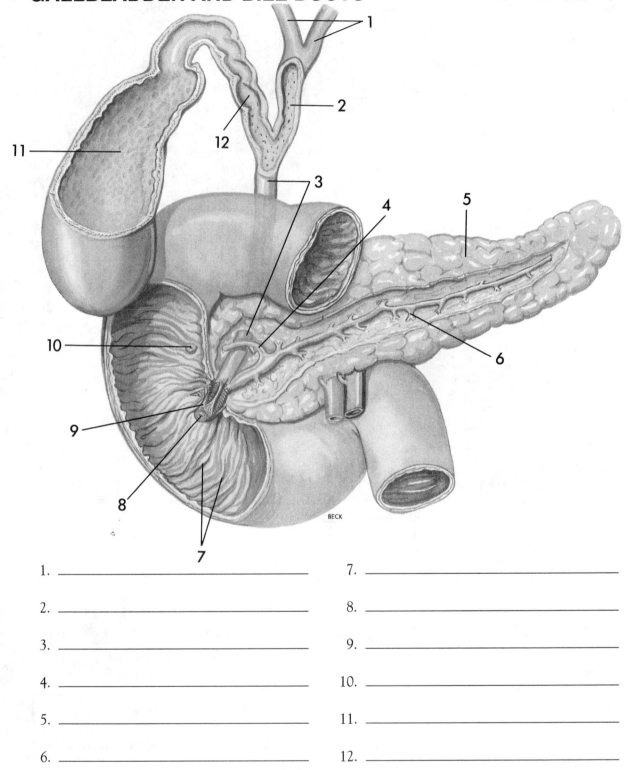

1. _____

2. _____

3. _____

4. _____

5. _____

6. _____

7. _____

8. _____

9. _____

10. _____

11. _____

12. _____

THE SMALL INTESTINE

SEGMENT OF JEJUNUM

THREE-DIMENSIONAL MAGNIFICATION
OF JEJUNAL WALL

THREE CELLS OF THE VILLUS' EPITHELIUM
SHOWING BRUSH BORDER
(MICROVILLI)

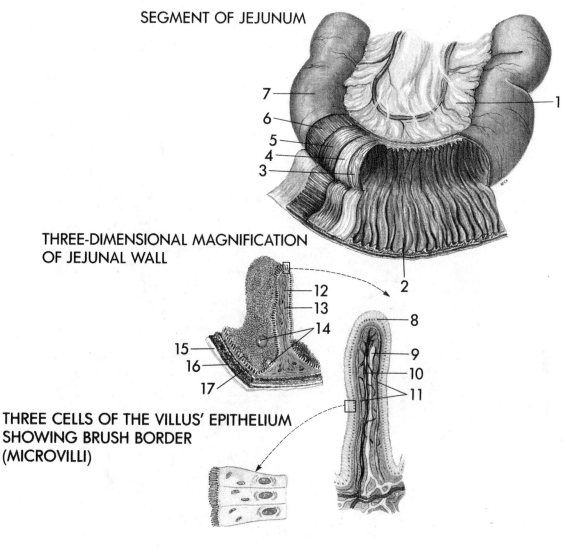

1. _____

2. _____

3. _____

4. _____

5. _____

6. _____

7. _____

8. _____

9. _____

10. _____

11. _____

12. _____

13. _____

14. _____

15. _____

16. _____

17. _____

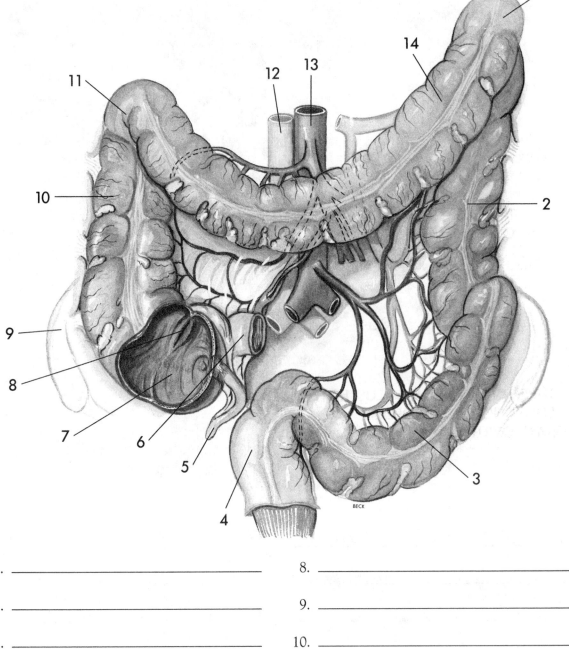

BECK

1. _____   8. _____

2. _____   9. _____

3. _____   10. _____

4. _____   11. _____

5. _____   12. _____

6. _____   13. _____

7. _____   14. _____

15 **Nutrition and Metabolism**

Most of us love to eat, but do the foods we enjoy provide us with the basic food types necessary for good nutrition? The body, a finely tuned machine, requires a balance of carbohydrates, fats, proteins, vitamins, and minerals to function properly. These nutrients must be digested, absorbed, and circulated to cells constantly to accommodate the numerous activities that occur throughout the body. The use the body makes of foods once these processes are completed is called "metabolism."

The liver plays a major role in the metabolism of food. It helps maintain a normal blood glucose level, removes toxins from the blood, processes blood immediately after it leaves the gastrointestinal tract, and initiates the first steps of protein and fat metabolism.

This chapter also discusses basal metabolic rate (BMR). The BMR is the rate at which food is catabolized under basal conditions. This test and the protein-bound iodine (PBI) are indirect measures of thyroid gland functioning. The total metabolic rate (TMR) is the amount of energy, expressed in calories, used by the body each day.

Finally, maintaining a constant body temperature is a function of the hypothalamus and a challenge for the metabolic factors of the body. Review of this chapter is necessary to provide you with an understanding of the "fuel" or nutrition necessary to maintain your complex homeostatic machine—the body.

TOPICS FOR REVIEW

Before progressing to Chapter 16, you should be able to define and contrast catabolism and anabolism. Your review should include the metabolic roles of carbohydrates, fats, proteins, vitamins, and minerals. Your study should conclude with an understanding of the basal metabolic rate and physiological mechanisms that regulate body temperature.

THE ROLE OF THE LIVER

Fill in the blanks.

The liver plays an important role in the mechanical digestion of lipids because it secretes (1) _____. It also produces two of the plasma proteins that play an essential role in blood clotting. These two proteins are (2) _____ and (3) _____. Additionally, liver cells store several substances, notably vitamins A, D, and (4) _____.

Finally, the liver is assisted by a unique structural feature of the blood vessels that supply it. This arrangement, known as the (5) _____ _____ _____, allows toxins to be removed from the bloodstream before nutrients are distributed throughout the body.

NUTRIENT METABOLISM

Match the term with the definition.

    (a) Carbohydrate   (d) Vitamins
    (b) Fat   (e) Minerals
    (c) Protein

_____ 6. Used if cells have inadequate amounts of glucose to catabolize
_____ 7. Preferred energy food
_____ 8. Amino acids
_____ 9. Fat soluble
_____ 10. Required for nerve conduction
_____ 11. Glycolysis
_____ 12. Inorganic elements found naturally in the earth
_____ 13. Pyruvic acid

Circle the word or phrase that does not belong.

14. Glycolysis	Citric acid cycle	ATP	Bile
15. Adipose	Amino acids	Triglycerides	Glycerol
16. A	D	M	K
17. Iron	Proteins	Amino acids	Essential
18. Hydrocortisone	Insulin	Growth hormone	Epinephrine
19. Sodium	Calcium	Zinc	Folic acid
20. Thiamine	Niacin	Ascorbic acid	Riboflavin

▶ *If you have had difficulty with this section, review pages 329-334.*

METABOLIC RATES
BODY TEMPERATURE

Circle the correct choice.

21. The rate at which food is catabolized under basal conditions is the:
    A. TMR   C. BMR
    B. PBI   D. ATP

22. The total amount of energy used by the body per day is the:
    A. TMR   C. BMR
    B. PBI   D. ATP

23. Over _____ of the energy released from food molecules during catabolism is converted to heat rather than being transferred to ATP.
    A. 20%   C. 60%
    B. 40%   D. 80%

24. Maintaining thermoregulation is a function of the:
    A. Thalamus   C. Thyroid
    B. Hypothalamus   D. Parathyroids

25. Transfer of heat energy to the skin, then the external environment is known as:
    A. Radiation   C. Convection
    B. Conduction   D. Evaporation

26. A flow of heat waves away from the blood is known as:
    A. Radiation   C. Convection
    B. Conduction   D. Evaporation

27. A transfer of heat energy to air that is continually flowing away from the skin is known as:
    A. Radiation   C. Convection
    B. Conduction   D. Evaporation

28. Heat that is absorbed by the process of water vaporization is called:
    A. Radiation   C. Convection
    B. Conduction   D. Evaporation

29. A/an _____ is the amount of energy needed to raise the temperature of one gram of water one degree Celsius.
    A. Calorie   C. ATP
    B. Kilocalorie   D. BMR

▷ *If you have had difficulty with this section, review pages 334-337.*

Unscramble the words.

30. LRIEV
    ⬜⬜⭕⭕⭕

31. TAOBALICMS
    ⬜⭕⭕⬜⬜⭕⬜⬜⬜⬜

32. OMNIA
    ⭕⬜⬜⭕⭕

33. YPURCVI
    ⭕⬜⬜⬜⬜⭕⬜

Take the circled letters, unscramble them, and fill in the statement.
How the magician paid his bills.

34. ⬜⬜⬜⬜⬜⬜⬜⬜⬜⬜⬜⬜⬜

APPLYING WHAT YOU KNOW

35. Dr. Carey was concerned about Deborrah. Her daily food intake provided fewer calories than her TMR. If this trend continues, what will be the result? If it continues over a long period of time, what eating disorder might Deborrah develop?

36. Mrs. Arce was experiencing fatigue and a blood test revealed that she was slightly anemic. What mineral will her doctor most likely prescribe? What dietary sources might you suggest that she emphasize in her daily intake?

37. Mr. Mobley was training daily for an upcoming marathon. Three days before the 25 mile event, he suddenly quit his daily routine of jogging and switched to a diet high in carbohydrates. Why did Mr. Mobley suddenly switch his routine of training?

38. WORD FIND

Can you find 18 terms from this chapter in the box of letters? Words may be spelled top to bottom, bottom to top, right to left, left to right, or diagonally.

```
C  C  C  B  W  E  F  F  L  J  V  G  G  S
A  T  N  L  W  E  U  O  Z  I  E  L  I  B
R  K  K  P  Z  F  R  I  T  P  O  Y  K  L
B  Q  M  I  N  E  R  A  L  S  J  C  C  P
O  S  S  N  C  X  M  I  D  B  D  O  W  P
H  S  I  Y  O  I  T  X  H  I  N  L  S  N
Y  E  L  N  N  I  K  W  W  D  S  Y  N  S
D  G  O  S  O  W  T  I  U  E  H  S  C  N
R  M  B  Y  T  I  W  C  Z  N  N  I  A  A
A  R  A  Q  W  A  T  B  E  I  G  S  J  M
T  E  T  F  N  I  F  A  E  V  E  W  T  Y
E  V  A  P  O  R  A  T  I  O  N  D  T  E
S  I  C  N  E  S  O  P  I  D  A  O  F  E
H  L  Y  K  I  R  E  B  V  P  A  H  C  J
I  W  E  E  P  A  D  F  T  E  A  R  G  G
```

ATP	Conduction	Liver
Adipose	Convection	Minerals
BMR	Evaporation	Proteins
Bile	Fats	Radiation
Carbohydrates	Glycerol	TMR
Catabolism	Glycolysis	Vitamins

DID YOU KNOW?

The amount of energy required to raise a 200-pound man 15 feet is about the amount of energy in one large calorie.

NUTRITION/METABOLISM

ACROSS

1. Breaks food molecules down releasing stored energy
4. Amount of energy needed to raise the temperature of one gram of water one degree Celsius
7. Rate of metabolism when a person is lying down, but awake (abbreviation)
8. A series of reactions that join glucose molecules together to form glycogen
10. Builds food molecules into complex substances

DOWN

2. Occurs when food molecules enter cells and undergo many chemical changes there
3. Organic molecule needed in small quantities for normal metabolism throughout the body
5. Oxygen-using
6. A unit of measure for heat, also known as a large calorie
9. Takes place in the cytoplasm of a cell and changes glucose to pyruvic acid

CHAPTER 16

The Urinary System

Living produces wastes. Wherever people live or work or play, wastes accumulate. To keep these areas healthy, there must be a method of disposing of these wastes such as a sanitation department.

Wastes accumulate in your body also. The conversion of food and gases into substances and energy necessary for survival results in waste products. A large percentage of these wastes is removed by the urinary system.

Two vital organs, the kidneys, cleanse the blood of the many waste products that are continually produced as a result of the metabolism of food in the body cells. They eliminate these wastes in the form of urine.

Urine formation is the result of three processes: filtration, reabsorption, and secretion. These processes occur in successive portions of the microscopic units of the kidneys known as nephrons. The amount of urine produced by the nephrons is controlled primarily by the hormones ADH and aldosterone.

After urine is produced it is drained from the renal pelvis by the ureters to flow into the bladder. The bladder then stores the urine until it is voided through the urethra.

If waste products are allowed to accumulate in the body, they soon become poisonous—a condition called *uremia*. A knowledge of the urinary system is necessary to understand how the body rids itself of waste and avoids toxicity.

TOPICS FOR REVIEW

Before progressing to Chapter 17, you should have an understanding of the structure and function of the organs of the urinary system. Your review should include knowledge of the nephron and its role in urine production. Your study should conclude with a review of the three main processes involved in urine production and the mechanisms that control urine volume.

KIDNEYS

Circle the correct choice.

1. The outermost portion of the kidney is known as the:
   A. Medulla
   B. Papilla
   C. Pelvis
   D. Pyramid
   E. Cortex

2. The saclike structure that surrounds the glomerulus is the:
   A. Renal pelvis
   B. Calyx
   C. Bowman's capsule
   D. Cortex
   E. None of the above is correct

3. The renal corpuscle is made up of the:
   A. Bowman's capsule and proximal convoluted tubule
   B. Glomerulus and proximal convoluted tubule
   C. Bowman's capsule and the distal convoluted tubule
   D. Glomerulus and the distal convoluted tubule
   E. Bowman's capsule and the glomerulus

4. Which of the following functions is *not* performed by the kidneys?
   A. Help maintain homeostasis
   B. Remove wastes from the blood
   C. Produce ADH
   D. Remove electrolytes from the blood

5. _____% of the glomerular filtrate is reabsorbed.
   A. 20
   B. 40
   C. 75
   D. 85
   E. 99

6. The glomerular filtration rate is _____ ml per minute.
   A. 1.25
   B. 12.5
   C. 125.0
   D. 1250.0
   E. None of the above is correct

7. Glucose is reabsorbed in the:
   A. Loop of Henle
   B. Proximal convoluted tubule
   C. Distal convoluted tubule
   D. Glomerulus
   E. None of the above is correct

8. Reabsorption does *not* occur in the:
   A. Loop of Henle
   B. Proximal convoluted tubule
   C. Distal convoluted tubule
   D. Collecting tubules
   E. Calyx

9. The greater the amount of salt intake the:
   A. Less salt excreted in the urine
   B. More salt is reabsorbed
   C. The more salt excreted in the urine
   D. None of the above is correct

10. Which one of the following substances is secreted by diffusion:
    A. Sodium ions
    B. Certain drugs
    C. Ammonia
    D. Hydrogen ions
    E. Potassium ions

11. Which of the following statements about ADH is *not* correct?
    A. It is stored by the pituitary gland
    B. It makes the collecting tubules less permeable to water
    C. It makes the distal convoluted tubules more permeable
    D. It is produced by the hypothalamus
12. Which of the following statements about aldosterone is *not* correct?
    A. It is secreted by the adrenal cortex
    B. It is a water-retaining hormone
    C. It is a salt-retaining hormone
    D. All of the above are correct

Choose the correct term and write its letter in the space next to the appropriate definition below.

a.	Medulla	h.	Uremia
b.	Cortex	i.	Proteinuria
c.	Pyramids	j.	Bowman's capsule
d.	Papilla	k.	Glomerulus
e.	Pelvis	l.	Loop of Henle
f.	Calyx	m.	CAPD
g.	Nephrons	n.	Glycosuria

_____ 13. Functioning unit of urinary system
_____ 14. Abnormally large amounts of plasma proteins in the urine
_____ 15. Uremic poisoning
_____ 16. Outer part of kidney
_____ 17. Together with Bowman's capsule forms renal corpuscle
_____ 18. Division of the renal pelvis
_____ 19. Cup-shaped top of a nephron
_____ 20. Innermost end of a pyramid
_____ 21. Extension of proximal tubule
_____ 22. Triangular-shaped divisions of the medulla of the kidney
_____ 23. Used in the treatment of renal failure
_____ 24. Inner portion of kidney

▷ *If you have had difficulty with this section, review pages 341-348 and 351.*

URETERS
URINARY BLADDER
URETHRA

Indicate which organ is identified by the following descriptions by inserting the appropriate letter in the answer blank.

       (a) Ureters   (b) Bladder   (c) Urethra

_____ 25. Rugae

_____ 26. Lower-most part of urinary tract

_____ 27. Lining membrane richly supplied with sensory nerve endings

_____ 28. Lies behind pubic symphysis

_____ 29. Dual function in male

_____ 30. 1 1/2 inches long in female

_____ 31. Drains renal pelvis

_____ 32. Surrounded by prostate in male

_____ 33. Elastic fibers and involuntary muscle fibers

_____ 34. 10 to 12 inches long

_____ 35. Trigone

Fill in the blanks.

36. _____ _____ is the description of the pain caused by the passage of a kidney stone.

37. The urinary tract is lined with _____ - _____.

38. Another name for kidney stones is _____ - _____.

39. A technique that uses _____ to pulverize stones, thus avoiding surgery, is being used to treat kidney stones.

40. The passage of a tube through the urethra into the bladder for the removal of urine is known as
_____.

41. The _____ - _____ is the basinlike upper end of the ureter located inside the kidney.

42. In the male, the urethra serves a dual function: a passageway for urine and _____.

43. The external opening of the urethra is the _____ - _____.

▶ *If you have had difficulty with this section, review pages 349-353.*

MICTURITION

Fill in the blanks.

The terms (44) _____, (45) _____, and (46) _____ all refer to the passage of urine from the body or the emptying of the bladder. The sphincters guard the bladder. The (47) _____ _____ sphincter is located at the bladder (48) _____ and is involuntary. The external urethral sphincter circles the (49) _____ and is under (50) _____ control. As the bladder fills, nervous impulses are transmitted to the spinal cord and an (51) _____ _____ is initiated. Urine then enters the (52) _____ to be eliminated. Urinary (53) _____ is a condition in which no urine is voided. Urinary (54) _____ is when the kidneys do not produce any urine, but the bladder retains its ability to empty itself. Complete destruction or transection of the sacral cord produces an (55) _____ _____.

▷ *If you have had difficulty with this section, review pages 353-354.*

APPLYING WHAT YOU KNOW

56. John suffered from low levels of ADH. What primary urinary symptom would he notice?
57. Bud was in a diving accident and his spinal cord was severed. He was paralyzed from the waist down and as a result was incontinent. His physician was concerned about the continuous residual urine buildup. What was the reason for concern?
58. Mrs. Lynch had a prolonged surgical procedure and experienced problems with urinary retention postoperatively. A urinary catheter was inserted into her bladder for the elimination of urine. Several days later Mrs. Lynch developed cystitis. What might be a possible cause?

59. WORD FIND

Can you find 18 terms from the chapter in the box of letters? Words may be spelled top to bottom, bottom to top, right to left, left to right, or diagonally.

```
C  C  C  B  W  E  F  F  L  J  V  G  G  S
A  T  N  L  W  E  U  O  Z  I  E  L  I  B
R  K  K  P  Z  F  R  I  T  P  O  Y  K  L
B  Q  M  I  N  E  R  A  L  S  J  C  C  P
O  S  S  N  C  X  M  I  D  B  D  O  W  P
H  S  I  Y  O  I  T  X  H  I  N  L  S  N
Y  E  L  N  N  I  K  W  W  D  S  Y  N  S
D  G  O  S  O  W  T  I  U  E  H  S  C  N
R  M  B  Y  T  I  W  C  Z  N  N  I  A  A
A  R  A  Q  W  A  T  B  E  I  G  S  J  M
T  E  T  F  N  I  F  A  E  V  E  W  T  Y
E  V  A  P  O  R  A  T  I  O  N  D  T  E
S  I  C  N  E  S  O  P  I  D  A  O  F  E
H  L  Y  K  I  R  E  B  V  P  A  H  C  J
I  W  E  E  P  A  D  F  T  E  A  R  G  G
```

ATP	Conduction	Liver
Adipose	Convection	Minerals
BMR	Evaporation	Proteins
Bile	Fats	Radiation
Carbohydrates	Glycerol	TMR
Catabolism	Glycolysis	Vitamins

DID YOU KNOW?

If the tubules in a kidney were stretched out and untangled, there would be 70 miles of them.

URINARY SYSTEM

ACROSS

3. Bladder infection
7. Absence of urine
8. Passage of a tube into the bladder to withdraw urine
11. Network of blood capillaries tucked into Bowman's capsule

DOWN

1. Urination
2. Ultrasound generator used to break up kidney stones
3. Division of the renal pelvis
4. Voiding involuntarily
5. Area on posterior bladder wall free of rugae
6. Glucose in the urine
9. Large amount of urine
10. Scanty urine

URINARY SYSTEM

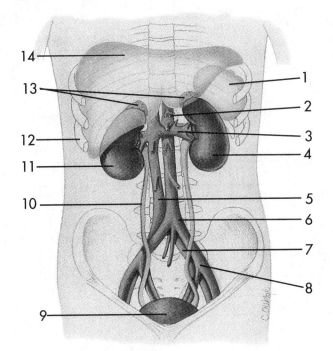

1. _____

2. _____

3. _____

4. _____

5. _____

6. _____

7. _____

8. _____

9. _____

10. _____

11. _____

12. _____

13. _____

14. _____

KIDNEY

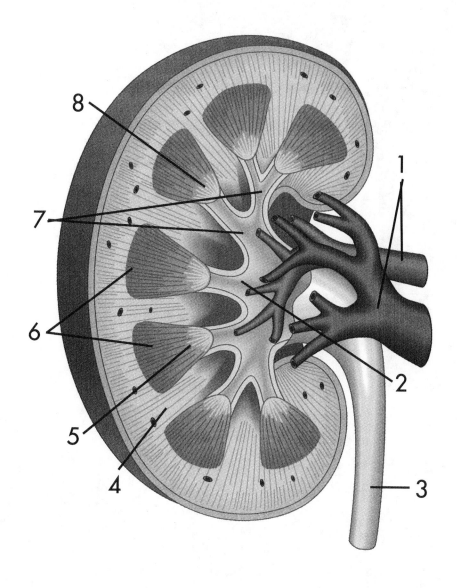

1. _____   5. _____

2. _____   6. _____

3. _____   7. _____

4. _____   8. _____

NEPHRON

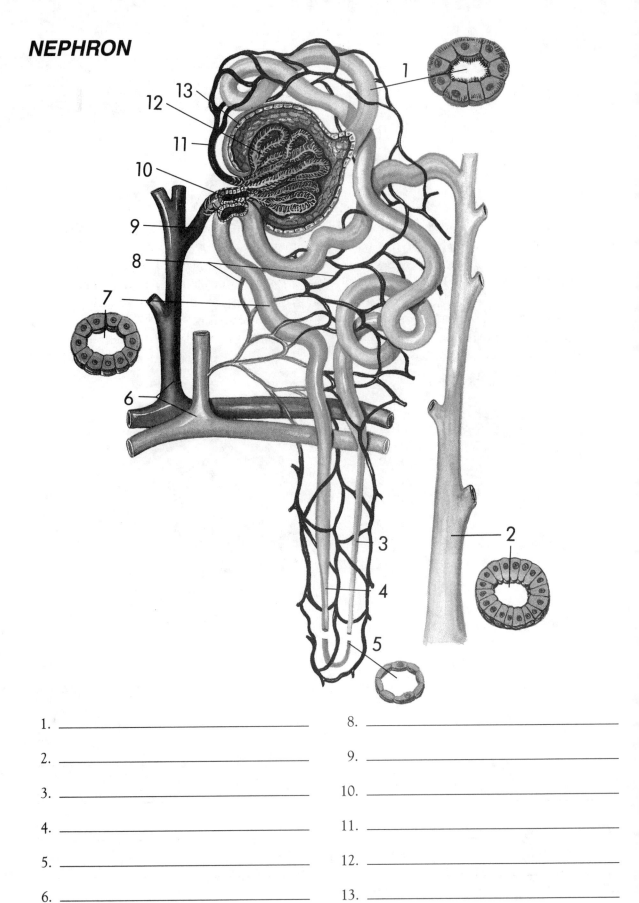

1. _____

2. _____

3. _____

4. _____

5. _____

6. _____

7. _____

8. _____

9. _____

10. _____

11. _____

12. _____

13. _____

CHAPTER 17   Fluid and Electrolyte Balance

Referring to the very first chapter in your text, you will recall that survival depends on the body's ability to maintain or restore homeostasis. Specifically, homeostasis means that the body fluids remain constant within very narrow limits. These fluids are classified as either intracellular fluid (ICF) or extracellular fluid (ECF). As their names imply, intracellular fluid lies within the cells and extracellular fluid is located outside the cells. A balance between these two fluids is maintained by certain body mechanisms. They are: (a) the adjustment of fluid output to fluid intake under normal circumstances; (b) the concentration of electrolytes in the extracellular fluid; (c) the capillary blood pressure, and finally (d) the concentration of proteins in the blood.

Comprehension of how these mechanisms maintain and restore fluid balance is necessary for an understanding of the complexities of homeostasis and its relationship to the survival of the individual.

TOPICS FOR REVIEW

Before progressing to Chapter 18, you should review the types of body fluids and their subdivisions. Your study should include the mechanisms that maintain fluid balance and the nature and importance of electrolytes in body fluids. You should be able to give examples of common fluid imbalances, and have an understanding of the role of fluid and electrolyte balance in the maintenance of homeostasis.

BODY FLUIDS

Circle the correct answer.
1. The largest volume of water by far lies (inside or outside) cells.
2. Interstitial fluid is (intracellular or extracellular).
3. Plasma is (intracellular or extracellular).
4. Obese people have a (lower or higher) water content per pound of body weight than thin people.
5. Infants have (more or less) water in comparison to body weight than adults of either sex.
6. There is a rapid (increase or decline) in the proportion of body water to body weight during the first year of life.
7. The female body contains slightly (more or less) water per pound of weight.
8. In general, as age increases, the amount of water per pound of body weight (increases or decreases).
9. Excluding adipose tissue, approximately (55% or 85%) of body weight is water.
10. The term (*fluid balance* or *fluid compartments*) means the volumes of ICF, IF, plasma, and the total volume of water in the body all remain relatively constant.

▶ *If you have had difficulty with this section, review pages 359-360.*

MECHANISMS THAT MAINTAIN FLUID BALANCE

Circle the correct choice.
11. Which one of the following is a positively charged ion?
    A. Chloride
    B. Calcium
    C. Sodium
    D. Potassium
12. Which one of the following is a negatively charged ion?
    A. Chloride
    B. Bicarbonate
    C. Phosphate
    D. Sodium
13. The most abundant electrolytes in the blood plasma are:
    A. NaCl
    B. KMg
    C. HCO3
    D. HPO4
    E. CaPO4
14. If the blood sodium concentration increases, then blood volume will:
    A. Increase
    B. Decrease
    C. Remain the same
    D. None of the above is correct
15. The smallest amount of water comes from:
    A. Water in foods that are eaten
    B. Ingested liquids
    C. Water formed from catabolism
    D. None of the above is correct
16. The greatest amount of water lost from the body is from the:
    A. Lungs
    B. Skin by diffusion
    C. Skin by sweat
    D. Feces
    E. Kidneys

17. Which one of the following is *not* a major factor that influences extracellular and intracellular fluid volumes?
    A. The concentration of electrolytes in the extracellular fluid
    B. The capillary blood pressure
    C. The concentration of proteins in blood
    D. All of the above are important factors
18. The type of fluid output that changes the most is:
    A. Water loss in the feces
    B. Water loss across the skin
    C. Water loss via the lungs
    D. Water loss in the urine
    E. None of the above is correct
19. The chief regulators of sodium within the body are the:
    A. Lungs
    B. Sweat glands
    C. Kidneys
    D. Large intestine
    E. None of the above is correct
20. Which of the following is *not* correct?
    A. Fluid output must equal fluid intake.
    B. ADH controls salt reabsorption in the kidney.
    C. Water follows sodium.
    D. Renal tubule regulation of salt and water is the most important factor in determining urine volume.
21. Diuretics work on all but which one of the following?
    A. Proximal tubule
    B. Loop of Henle
    C. Distal tubule
    D. Collecting ducts
    E. Diuretics work on all of the above
22. Of all the sodium-containing secretions, the one with the largest volume is:
    A. Saliva
    B. Gastric secretions
    C. Bile
    D. Pancreatic juice
    E. Intestinal secretions
23. The higher the capillary blood pressure, the _____ the amount of interstitial fluid.
    A. Smaller
    B. Larger
    C. There is no relationship between capillary blood pressure and volume of interstitial fluid
24. An increase in capillary blood pressure will lead to _____ in blood volume.
    A. An increase
    B. A decrease
    C. No change
    D. None of the above is correct
25. Which one of the fluid compartments varies the most in volume?
    A. Intracellular
    B. Interstitial
    C. Extracellular
    D. Plasma
26. Which one of the following will *not* cause edema?
    A. Retention of electrolytes in the extracellular fluid
    B. Increase in capillary blood pressure
    C. Burns
    D. Decrease in plasma proteins
    E. All of the above may cause edema

If the following statements are true, insert T in the answer blanks. If any of the statements are false, circle the incorrect word(s) and write the correct word in the answer blank.

_____ 27. The three sources of fluid intake are: the liquids we drink, the foods we eat, and water formed by the anabolism of foods.

_____ 28. The body maintains fluid balance mainly by changing the volume of urine excreted to match changes in the volume of fluid intake.

_____ 29. Some output of fluid will occur as long as life continues.

_____ 30. Glucose is an example of an electrolyte.

_____ 31. Where sodium goes, water soon follows.

_____ 32. Excess aldosterone leads to hypovolemia.

_____ 33. Diuretics have their effect on glomerular function.

_____ 34. Typical daily intake and output totals should be approximately 1200 ml.

_____ 35. Bile is a sodium-containing internal secretion.

_____ 36. The average daily diet contains about 500 mEq of sodium.

▶ *If you have had difficulty with this section, review pages 359-367.*

FLUID IMBALANCES

Fill in the blanks.

(37)_____ is the fluid imbalance seen most often. In this condition, interstitial fluid volume (38) _____ first, but eventually, if treatment has not been given, intracellular fluid and plasma volumes (39) _____. (40) _____ can also occur, but is much less common. Giving (41) _____ _____ too rapidly or in too large amounts can put too heavy a burden on the (42) _____.

▶ *If you have had difficulty with this section, review page 367.*

APPLYING WHAT YOU KNOW

43. Mrs. Titus was asked to keep an accurate record of her fluid intake and output. She was concerned because the two did not balance. What is a possible explanation for this?

44. Nurse Briker was caring for a patient who was receiving diuretics. What special nursing implications should be followed for patients on this therapy?

45. WORD FIND

Can you find the 12 terms from this chapter in the box of letters? Words may be spelled top to bottom, bottom to top, right to left, left to right, or diagonally.

```
S  T  V  H  O  M  E  O  S  T  A  S  I  S  H
I  E  D  E  M  A  S  Q  A  L  P  U  E  G  L
M  M  L  U  F  L  U  I  D  O  Y  O  I  S  L
W  S  B  E  A  E  C  O  L  D  P  N  I  E  R
L  I  T  A  C  V  S  H  Q  O  C  E  H  B  C
L  L  X  J  L  T  E  A  W  Y  B  V  Q  Q  O
I  O  O  P  E  A  R  U  C  T  C  A  A  R  I
N  B  H  R  X  S  N  O  I  I  P  R  T  C  N
S  A  O  X  M  G  J  C  L  N  J  T  B  A  H
I  N  X  X  D  V  U  D  E  Y  K  N  Y  O  C
E  A  A  J  Q  D  I  U  R  E  T  I  C  S  X
I  F  C  J  A  M  P  V  E  N  B  E  Y  F  W
V  F  K  T  Q  X  R  M  D  D  S  I  V  O  T
E  T  T  Z  N  T  R  P  X  I  M  L  J  F  I
S  W  Y  A  P  V  Q  N  S  K  T  K  W  B  P
```

Aldosterone	Edema	Imbalance
Anabolism	Electrolyte	Intravenous
Catabolism	Fluid	Ions
Diuretics	Homeostasis	Kidney

DID YOU KNOW?

The best fluid replacement drink is 1/4 teaspoon of table salt to one quart of water.

FLUID/ELECTROLYTES

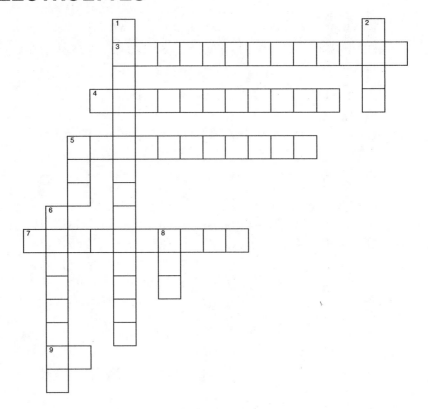

ACROSS

3. Result of rapidly given intravenous fluids

4. Result of large loss of body fluids

5. Compound that dissociates in solution into ions

7. To break up

9. A subdivision of extracellular fluid (abbreviation)

DOWN

1. Organic substance that doesn't dissociate in solution

2. Dissociated particles of an electrolyte that carry an electrical charge

5. Fluid outside cells (abbreviation)

6. "Causing urine"

8. Fluid inside cells (abbreviation)

CHAPTER **18** # Acid-Base Balance

It has been established in previous chapters that an equilibrium between intracellular and extracellular fluid volume must exist for homeostasis. Equally important to homeostasis is the chemical acid-base balance of the body fluids. The degree of acidity or alkalinity of a body fluid is expressed in pH value. The neutral point, where a fluid would be neither acid nor alkaline, is pH 7. Increasing acidity is expressed as less than 7, and increasing alkalinity as greater than 7. Examples of body fluids that are acidic are gastric juice (1.6) and urine (6.0). Blood, on the other hand, is considered alkaline with a pH of 7.45.

Buffers are substances that prevent a sharp change in the pH of a fluid when an acid or base is added to it. They are one of several mechanisms that are constantly monitoring the pH of fluids in the body. If, for any reason, these mechanisms do not function properly, a pH imbalance occurs. These two kinds of imbalances are known as alkalosis and acidosis.

Maintaining the acid-base balance of body fluids is a matter of vital importance. If this balance varies even slightly, necessary chemical and cellular reactions cannot occur. Your review of this chapter is necessary to understand the delicate fluid balance necessary to survival.

TOPICS FOR REVIEW

Before progressing to Chapter 19, you should have an understanding of the pH of body fluids and the mechanisms that control the pH of these fluids in the body. Your study should conclude with a review of the metabolic and respiratory types of pH imbalances.

pH OF BODY

Write the letter of the correct term on the blank next to the appropriate statement.

        (a) Acid   (b) Base

_____ 1. Lower concentration of hydrogen ions than hydroxide ions

_____ 2. Higher concentration of hydrogen ions than hydroxide ions

_____ 3. Gastric juice

_____ 4. Saliva

_____ 5. Arterial blood

_____ 6. Venous blood

_____ 7. Baking soda

_____ 8. Milk

_____ 9. Ammonia

_____ 10. Egg white

▶ *If you have had difficulty with this section, review pages 371-372.*

MECHANISMS THAT CONTROL pH OF BODY FLUIDS

Circle the correct choice.

11. When carbon dioxide enters the blood it reacts with the enzyme carbonic anhydrase to form:

    A. Sodium bicarbonate   D. Bicarbonate ion

    B. Water and carbon dioxide   E. Carbonic acid

    C. Ammonium chloride

12. The lungs remove _____ liters of carbonic acid each day.

    A. 10.0   D. 25.0

    B. 15.0   E. 30.0

    C. 20.0

13. When a buffer reacts with a strong acid it changes the strong acid to a:

    A. Weak acid   D. Water

    B. Strong base   E. None of the above is correct

    C. Weak base

14. Which one of the following is *not* a change in the blood that results from the buffering of fixed acids in tissue capillaries?

    A. The amount of carbonic acid increases slightly.

    B. The amount of bicarbonate in blood decreases.

    C. The hydrogen ion concentration of blood increases slightly.

    D. The blood pH decreases slightly.

    E. All of the above are changes that result from the buffering of fixed acids in tissue capillaries.

15. The most abundant acid in the body is:

    A. HCl   D. Acetic acid

    B. Lactic acid   E. Sulfuric acid

    C. Carbonic acid

16. The normal ratio of sodium bicarbonate to carbonic acid in arterial blood is:
   A. 5:1
   B. 10:1
   C. 15:1
   D. 20:1
   E. None of the above is correct
17. Which of the following would *not* be a consequence of holding your breath?
   A. The amount of carbonic acid in the blood would increase.
   B. The blood pH would decrease.
   C. The body would develop an alkalosis.
   D. No carbon dioxide could leave the body.
18. Which of the following is *not* true of the kidneys?
   A. They can eliminate larger amounts of acid than the lungs.
   B. More bases than acids are usually excreted by the kidneys.
   C. If the kidneys fail, homeostasis of acid-base balance fails.
   D. They are the most effective regulators of blood pH.
19. The pH of the urine may be as low as:
   A. 1.6
   B. 2.5
   C. 3.2
   D. 4.8
   E. 7.4
20. In the distal tubule cells the product of the reaction aided by carbonic anhydrase is:
   A. Water
   B. Carbon dioxide
   C. Water and carbon dioxide
   D. Hydrogen ions
   E. Carbonic acid
21. In the distal tubule, _____ leaves the tubule cells and enters the blood capillaries.
   A. Carbon dioxide
   B. Water
   C. HCO_3
   D. NaH_2PO_4
   E. $NaHCO_3$

Mark T in the answer blank if the statement is true. If the statement is false, circle the incorrect word(s) and correct the statement on the answer blank.

_____ 22. The body has three mechanisms for regulating the pH of its fluids: the heart mechanism, the respiratory mechanism, and the urinary mechanism.

_____ 23. Buffers consist of two kinds of substances and are therefore often called duobuffers.

_____ 24. More acids than bases are usually added to body fluids.

_____ 25. Some athletes have adopted a technique called bicarbonate loading, ingesting large amounts of sodium bicarbonate ($NaHCO_3$) to counteract the effects of lactic acid buildup.

_____ 26. Anything that causes an excessive increase in respiration will in time produce acidosis.

_____ 27. The lungs are the body's most effective regulator of blood pH.

_____ 28. More acids than bases are usually excreted by the kidneys because more acids than bases usually enter the blood.

_____ 29. Blood levels of sodium bicarbonate can be regulated by the lungs.

_____ 30. Blood levels of carbonic acid can be regulated by the kidneys.

▶ *If you have had difficulty with this section, review pages 371-377.*

METABOLIC AND RESPIRATORY DISTURBANCES

Write the letter of the correct term on the blank next to the appropriate definition.

a. Metabolic acidosis
b. Metabolic alkalosis
c. Respiratory acidosis
d. Respiratory alkalosis
e. Vomiting

f. Normal saline
g. Uncompensated metabolic acidosis
h. Hyperventilation
i. Hypersalivation
j. Ipecac

_____ 31. Emesis
_____ 32. Result of untreated diabetes
_____ 33. Chloride-containing solution
_____ 34. Bicarbonate deficit
_____ 35. Present during emesis
_____ 36. Bicarbonate excess
_____ 37. Rapid breathing
_____ 38. Carbonic acid excess
_____ 39. Carbonic acid deficit
_____ 40. Emetic

▶ *If you have had difficulty with this section, review pages 377-379.*

APPLYING WHAT YOU KNOW

41. Holly was pregnant and was experiencing repeated vomiting episodes for several days. Her doctor became concerned, admitted her to the hospital, and began intravenous administrations of normal saline. How will this help Holly?

42. Cara had a minor bladder infection. She had heard that this is often the result of the urine being less acidic than necessary, and that she should drink cranberry juice to correct the acid problem. She had no cranberry juice, so she decided to substitute orange juice. What was wrong with this substitution?

43. Mr. Cameron has frequent bouts of hyperacidity of the stomach. Which will assist in neutralizing the acid more promptly: milk or milk of magnesia?

44. WORD FIND

Can you find 18 terms from this chapter in the box of letters? Words may be spelled top to bottom, bottom to top, right to left, left to right, or diagonally.

```
S  I  S  A  T  S  O  E  M  O  H  P  R  G
E  C  N  A  L  A  B  D  I  U  L  F  E  F
T  S  Y  E  N  D  I  K  W  T  I  K  A  J
Y  D  M  N  D  E  J  W  L  P  T  D  J  I
L  D  N  O  L  H  V  W  O  U  H  D  O  I
O  V  E  R  H  Y  D  R  A  T  I  O  N  S
R  E  I  E  E  D  E  M  A  U  R  O  R  Q
T  L  M  T  C  R  U  L  R  F  S  E  O  Y
C  C  U  S  Z  A  W  E  K  A  T  N  I  X
E  O  Z  O  T  T  A  N  A  U  N  M  F
L  F  K  D  P  I  O  I  W  W  Q  L  G  Q
E  N  I  L  C  O  O  H  O  W  S  B  X  S
N  J  X  A  L  N  F  S  U  N  L  J  J  J
O  C  U  V  S  A  L  G  T  I  S  C  Z  X
N  I  Z  L  L  D  Y  X  Q  Q  K  D  C  D
```

ADH	Edema	Nonelectrolytes
Aldosterone	Electrolytes	Output
Anions	Fluid balance	Overhydration
Cations	Homeostasis	Sodium
Dehydration	Intake	Thirst
Diuretic	Kidneys	Water

DID YOU KNOW?

The brain is a 3-lb, greedy organ that demands 17% of all cardiac output and 20% of all available oxygen.

ACID/BASE BALANCE

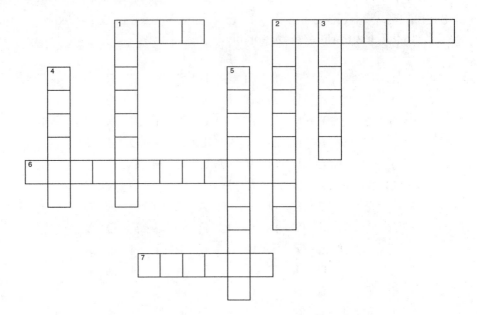

ACROSS

1. Substance with a pH lower than 7.0
2. Acid-base imbalance
6. Results from the excessive metabolism of fats in uncontrolled diabetics (2 words)
7. Vomitus

DOWN

1. Substance with a pH higher than 7.0
2. Serious complication of vomiting
3. Emetic
4. Prevents a sharp change in the pH of fluids
5. Released as a waste product from working muscles (2 words)

CHAPTER **19** # The Reproductive Systems

The reproductive system consists of those organs that participate in perpetuating the species. It is a unique body system in that its organs differ between the two sexes, and yet the goal of creating a new being is the same. Of interest also is the fact that this system is the only one not necessary to the survival of the individual, and yet survival of the species depends on the proper functioning of the reproductive organs. The male reproductive system is divided into the external genitals, the testes, the duct system, and accessory glands. The testes, or gonads, are considered essential organs because they produce the sex cells, sperm, which join with the female sex cells, ova, to form a new human being. They also secrete the male sex hormone, testosterone, which is responsible for the physical transformation of a boy to a man.

Sperm are formed in the testes by the seminiferous tubules. From there they enter a long narrow duct, the epididymis. They continue onward through the vas deferens into the ejaculatory duct, down the urethra, and out of the body. Throughout this journey, various glands secrete substances that add motility to the sperm and create a chemical environment conducive to reproduction.

The female reproductive system is truly extraordinary and diverse. It produces ova, receives the penis and sperm during intercourse, is the site of conception, houses and nourishes the embryo during prenatal development, and nourishes the infant after birth.

Because of its diversity, the physiology of the female is generally considered to be more complex than that of the male. Much of the activity of this system revolves around the menstrual cycle and the monthly preparation that the female undergoes for a possible pregnancy.

The organs of this system are divided into essential organs and accessory organs of reproduction. The essential organs of the female are the ovaries. Just as with the male, the essential organs of the female are referred to as the gonads. The gonads of both sexes produce the sex cells. In the male, the gonads produce the sperm and in the female they produce the ova. The gonads are also responsible for producing the hormones in each sex necessary for the appearance of the secondary sex characteristics.

The menstrual cycle of the female typically covers a period of 28 days. Each cycle consists of three phases: the menstrual period, the postmenstrual phase, and the premenstrual phase. Changes in the blood levels of the hormones that are responsible for the menstrual cycle also cause physical and emotional changes in the female. Knowledge of these phenomena and this system, in both the male and the female, is necessary to complete your understanding of the reproductive system.

TOPICS FOR REVIEW

Before progressing to Chapter 20, you should familiarize yourself with the structure and function of the organs of the male and female reproductive systems. Your review should include emphasis on the gross and microscopic structure of the testes and the production of sperm and testosterone. Your study should continue by tracing the pathway of a sperm cell from formation to expulsion from the body.

You should then familiarize yourself with the structure and function of the organs of the female reproductive system. Your review should include emphasis on the development of a mature ova from ovarian follicles, and should additionally concentrate on the phases and occurrences in a typical 28-day menstrual cycle.

MALE REPRODUCTIVE SYSTEM STRUCTURAL PLAN

Match the term on the left with the proper selection on the right.

Group A

_____	1. Testes	A. Fertilized ovum
_____	2. Spermatozoa	B. Accessory organ
_____	3. Ova	C. Male sex cell
_____	4. Penis	D. Gonads
_____	5. Zygote	E. Gamete

Group B

_____	6. Testes	A. Cowper's gland
_____	7. Bulbourethral	B. Scrotum
_____	8. Asexual	C. Essential organ
_____	9. External genitalia	D. Single parent
_____	10. Prostate	E. Accessory organ

▷ *If you have had difficulty with this section, review pages 383-384.*

TESTES

Circle the correct choice.

11. The testes are surrounded by a tough membrane called:
    A. Ductus deferens   C. Septum
    B. Tunica albuginea   D. Seminiferous membrane
12. The _____ lie near the septa that separate the lobules.
    A. Ductus deferens   C. Interstitial cells
    B. Sperm   D. Nerves
13. Sperm are found in the walls of the _____.
    A. Seminiferous tubule   C. Septum
    B. Interstitial cells   D. Blood vessels

14. An undescended testicle is:
    A. Orchidalgia
    B. Orchidorrhaphy
    C. Orchichorea
    D. Cryptorchidism
15. The structure that produces testosterone is (are) the:
    A. Seminiferous tubules
    B. Prostate gland
    C. Bulbourethral gland
    D. Pituitary gland
    E. Interstitial cells
16. The part of the sperm that contains genetic information that will be inherited is the:
    A. Tail
    B. Neck
    C. Middle piece
    D. Head
    E. Acrosome
17. Which one of the following is *not* a function of testosterone?
    A. It causes a deepening of the voice.
    B. It promotes the development of the male accessory glands.
    C. It has a stimulatory effect on protein catabolism.
    D. It causes greater muscular development and strength.
18. Sperm production is called:
    A. Spermatogonia
    B. Spermatids
    C. Spermatogenesis
    D. Spermatocyte
19. The section of the sperm that contains enzymes that enable it to break down the covering of the ovum and permit entry should contact occur is the:
    A. Acrosome
    B. Midpiece
    C. Tail
    D. Stem
20. Descent of the testes usually occurs about _____.
    A. Two months after birth
    B. Two months before birth
    C. Two months after conception
    D. Two years after birth
    E. None of the above is correct

Fill in the blanks.

The (21) _____ are the gonads of the male. From puberty on, the seminiferous tubules are continuously forming (22) _____. Any of these cells may join with the female sex cell, the (23) _____, to become a new human being.

Another function of the testes is to secrete the male hormone (24) _____ that transforms a boy to a man. This hormone is secreted by the (25) _____ _____ of the testes. A good way to remember testosterone's functions is to think of it as "the (26) _____ hormone" and "the (27) _____ hormone."

▶ *If you have had difficulty with this section, review pages 384-389.*

REPRODUCTIVE DUCTS
ACCESSORY OR SUPPORTIVE SEX GLANDS
EXTERNAL GENITALIA

Choose the correct term and write its letter in the space next to the appropriate definition below.

a. Epididymis   f. Prostate gland
b. Vas deferens   g. Cowper's gland
c. Ejaculatory duct   h. Prostatectomy
d. Prepuce   i. Semen
e. Seminal vesicles   j. Scrotum

_____ 28. Continuation of ducts that start in epididymis
_____ 29. Procedure performed for benign prostatic hypertrophy
_____ 30. Also known as bulbourethral
_____ 31. Narrow tube that lies along the top and behind the testes
_____ 32. Doughnut-shaped gland beneath bladder
_____ 33. Continuation of vas deferens
_____ 34. Mixture of sperm and secretions of accessory sex glands
_____ 35. Contributes 60% of the seminal fluid volume
_____ 36. Removed during circumcision
_____ 37. External genitalia

▶ *If you have had difficulty with this section, review pages 389-391.*

FEMALE REPRODUCTIVE SYSTEM
STRUCTURAL PLAN

Match the term on the left with the proper selection on the right.

_____ 38. Ovaries   A. Genitals
_____ 39. Vagina   B. Accessory sex gland
_____ 40. Bartholin   C. Accessory duct
_____ 41. Vulva   D. Gonads
_____ 42. Ova   E. Sex cell

Write the letter of the correct description in the blank next to the appropriate structure.

(a) External structure   (b) Internal structure

_____ 43. Mons pubis   _____ 47. Vestibule
_____ 44. Vagina   _____ 48. Clitoris
_____ 45. Labia majora   _____ 49. Labia minora
_____ 46. Uterine tubes   _____ 50. Ovaries

▶ *If you have had difficulty with this section, review pages 391-397.*

OVARIES

Fill in the blanks.

The ovaries are the (51) _____ of the female. They have two main functions. The first is the production of the female sex cell. This process is called (52) _____. The specialized type of cell division that occurs during sexual cell reproduction is known as (53) _____. The ovum is the body's largest cell and has (54) _____ _____ the number of chromosomes found in other body cells. At the time of (55) _____, the sex cells from both parents fuse and (56) _____ chromosomes are united.

The second major function of the ovaries is to secrete the sex hormones (57) _____ and (58) _____. Estrogen is the sex hormone that causes the development and mainte-nance of the female (59) _____ _____ _____. Progesterone acts with estrogen to help initiate the (60) _____ _____ in girls entering (61) _____.

▶ *If you have had difficulty with this section, review pages 392-394.*

FEMALE REPRODUCTIVE DUCTS

Write the letter of the correct structure in the blank next to the appropriate definition.

    (a) Uterine tubes   (b) Uterus   (c) Vagina

_____ 62. Ectopic pregnancy
_____ 63. Lining known as endometrium
_____ 64. Terminal end of birth canal
_____ 65. Site of menstruation
_____ 66. Approximately 4 inches in length
_____ 67. Consists of body, fundus, and cervix
_____ 68. Site of fertilization
_____ 69. Also known as oviduct
_____ 70. Entrance way for sperm
_____ 71. Total hysterectomy

▶ *If you have had difficulty with this section, review pages 394-396 and 400.*

ACCESSORY OR SUPPORTIVE SEX GLANDS
EXTERNAL GENITALS OF THE FEMALE

Match the term on the left with the proper selection on the right.

Group A

_____ 72. Bartholin's gland

_____ 73. Breasts

_____ 74. Alveoli

_____ 75. Lactiferous ducts

_____ 76. Areola

A. Colored area around nipple
B. Grapelike clusters of milk-secreting cells
C. Drain alveoli
D. Secretes lubricating fluid
E. Primarily fat tissue

Group B

_____ 77. Mons pubis

_____ 78. Labia majora

_____ 79. Clitoris

_____ 80. Vestibule

_____ 81. Episiotomy

A. "Large lips"
B. Area between labia minora
C. Surgical procedure
D. Composed of erectile tissue
E. Pad of fat over the symphysis pubis

▷ *If you have had difficulty with this section, review pages 396-397.*

MENSTRUAL CYCLE

If the following statement is true, insert T in the answer blank. If the statement is false, circle the incorrect word(s) and insert the correct word(s) in the answer blank.

_____ 82. Climacteric is the scientific name for the beginning of the menses.

_____ 83. As a general rule, several ovum mature each month during the 30-40 years that a woman has menstrual periods.

_____ 84. Ovulation occurs 28 days before the next menstrual period begins.

_____ 85. The first day of ovulation is considered the first day of the cycle.

_____ 86. A woman's fertile period lasts only a few days out of each month.

_____ 87. The control of the menstrual cycle lies in the posterior pituitary gland.

Write the letter of the correct hormone in the blank next to the appropriate description.

    (a) FSH   (b) LH

_____ 88. Ovulating hormone

_____ 89. Secreted during first days of menstrual cycle

_____ 90. Secreted after estrogen level of blood increases

_____ 91. Causes final maturation of follicle and ovum

_____ 92. Birth control pills suppress this one

▷ *If you have had difficulty with this section, review pages 397-400.*

APPLYING WHAT YOU KNOW

93. Mr. Belinki is going into the hospital for the surgical removal of his testes. As a result of this surgery, will Mr. Belinki be impotent?
94. When baby Ross was born, the pediatrician discovered that his left testicle had not descended into the scrotum. If this situation is not corrected soon, might baby Ross be sterile or impotent?
95. Ms. Gaynes contracted gonorrhea. By the time she made an appointment to see her doctor, it had spread to her abdominal organs. How is this possible when gonorrhea is a disease of the reproductive system?
96. Mrs. Harlan was having a bilateral oophorectomy. Is this a sterilization procedure? Will she experience menopause?
97. Ms. Comstock had a total hysterectomy. Will she experience menopause?
98. WORD FIND

Can you find 18 terms from this chapter in the box of letters? Words may be spelled top to bottom, bottom to top, right to left, left to right, or diagonally.

```
M  K  O  V  I  D  U  C  T  S  E  D  H  G
S  E  I  R  A  V  O  I  F  U  T  G  L  L
I  N  H  Z  H  M  P  M  K  O  H  I  C  W
D  D  V  A  S  D  E  F  E  R  E  N  S  H
I  O  A  C  C  I  P  J  V  E  H  Y  D  S
H  M  G  R  R  B  M  N  X  F  G  Y  I  O
C  E  I  O  O  E  Z  Y  T  I  C  S  T  E
R  T  N  S  T  P  S  D  D  N  O  V  A  D
O  R  A  O  U  W  E  B  A  I  J  S  M  Z
T  I  C  M  M  E  Y  N  E  M  D  L  R  V
P  U  A  E  U  D  G  M  I  E  H  I  E  H
Y  M  O  T  C  E  T  A  T  S  O  R  P  B
R  P  E  E  R  M  N  E  G  O  R  T  S  E
C  O  W  P  E  R  S  I  N  X  K  E  A  P
```

Acrosome	Meiosis	Scrotum
Cowpers	Ovaries	Seminiferous
Cryptorchidism	Oviducts	Sperm
Endometrium	Penis	Spermatids
Epididymis	Pregnancy	Vagina
Estrogen	Prostatectomy	Vas deferens

DID YOU KNOW?

The testes produce approximately 50 million sperm per day. Every 2 to 3 months they produce enough cells to populate the entire earth.

REPRODUCTIVE SYSTEM

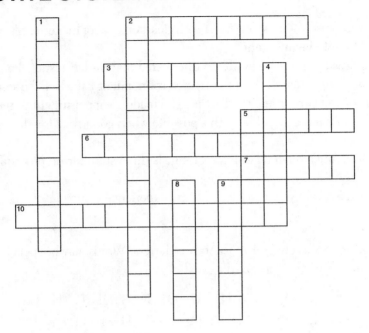

ACROSS

2. Female erectile tissue
3. Colored area around nipple
5. Male reproductive fluid
6. Sex cells
7. External genitalia
10. Male sex

DOWN

1. Failure to have a menstrual period
2. Surgical removal of foreskin
4. Foreskin
8. Menstrual period
9. Essential organs of reproduction

MALE REPRODUCTIVE ORGANS

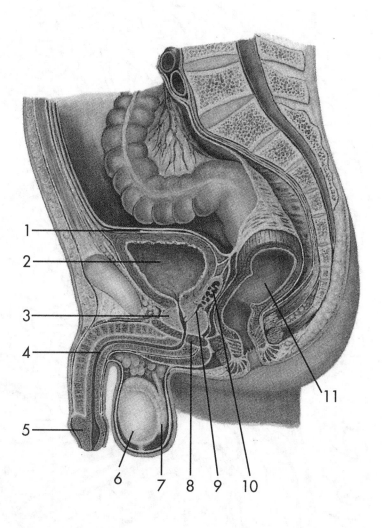

1. _____

2. _____

3. _____

4. _____

5. _____

6. _____

7. _____

8. _____

9. _____

10. _____

11. _____

TUBULES OF TESTIS AND EPIDIDYMIS

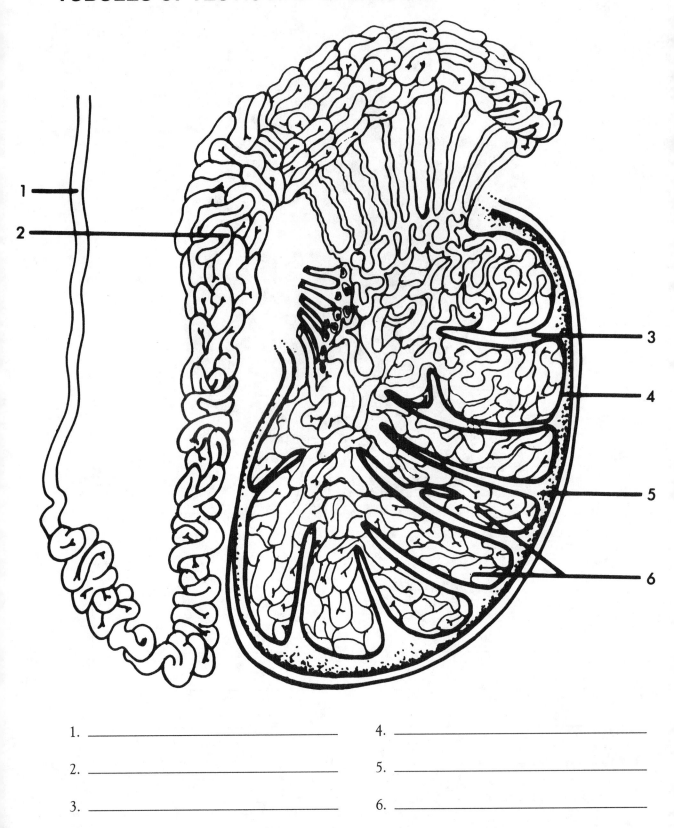

1. _____   4. _____

2. _____   5. _____

3. _____   6. _____

VULVA

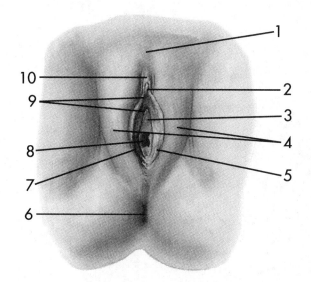

1. _____   6. _____

2. _____   7. _____

3. _____   8. _____

4. _____   9. _____

5. _____   10. _____

BREAST

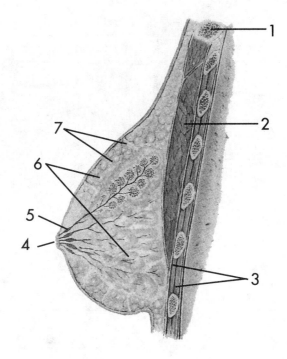

1. _____   5. _____

2. _____   6. _____

3. _____   7. _____

4. _____

FEMALE PELVIS

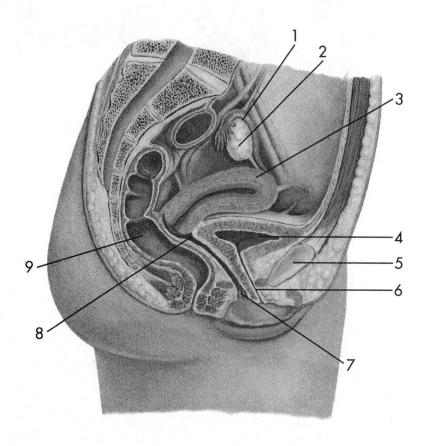

1. _____

2. _____

3. _____

4. _____

5. _____

6. _____

7. _____

8. _____

9. _____

UTERUS AND ADJACENT STRUCTURES

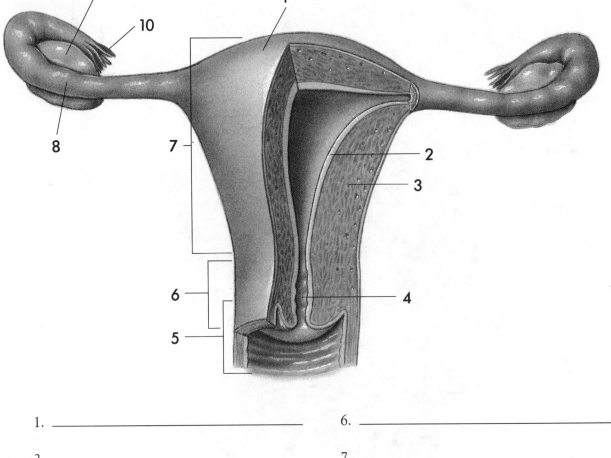

1. _____

2. _____

3. _____

4. _____

5. _____

6. _____

7. _____

8. _____

9. _____

10. _____

CHAPTER 20

Growth and Development

Millions of fragile microscopic sperm swim against numerous obstacles to reach the ova and create a new life. At birth, the newborn will fill his lungs with air and cry lustily, signaling to the world that he is ready to begin the cycle of life. This cycle will be marked by ongoing changes, periodic physical growth, and continuous development.

This chapter reviews the more significant events that occur in the normal growth and development of an individual from conception to death. Realizing that each individual is unique, we nonetheless can discover amid all the complexities of humanity some constants that are understandable and predictable.

A knowledge of human growth and development is essential in understanding the commonalities that influence individuals as they pass through the cycle of life.

TOPICS FOR REVIEW

Your review of this chapter should include an understanding of the concept of development as a biological process. You should familiarize yourself with the major developmental changes from conception through older adulthood. Your study should conclude with a review of the effects of aging on the body systems.

PRENATAL PERIOD

Fill in the blanks.

The prenatal stage of development begins at the time of (1) _____ and continues until (2) _____. The science of the development of an individual before birth is called (3) _____. Fertilization takes place in the outer third of the (4) _____. The fertilized ovum or (5) _____ begins to divide and in approximately 3 days forms a solid mass called a (6) _____. By the time it enters the uterus, it is a hollow ball of cells called a (7) _____. As it continues to develop, it forms a structure with two cavities. The (8) _____ _____ will become a fluid-filled sac for the embryo. The (9) _____ will develop into an important fetal membrane in the (10) _____.

Choose the correct term and write its letter in the space next to the appropriate definition below.

a. Laparoscope
b. Gestation
c. Antenatal
d. Histogenesis
e. Quickening
f. Endoderm
g. In vitro
h. Parturition
i. Embryonic phase
j. Ultrasonogram

_____ 11. "Within a glass"
_____ 12. Inside germ layer
_____ 13. Before birth
_____ 14. Length of pregnancy
_____ 15. Fiberoptic viewing instrument
_____ 16. Process of birth
_____ 17. First fetal movement
_____ 18. Study of how the primary germ layers develop into many different kinds of tissues
_____ 19. Fertilization until the end of the eighth week of gestation
_____ 20. Monitors progress of developing fetus

▶ *If you have had difficulty with this section, review pages 407-416.*

POSTNATAL PERIOD

Circle the correct choice.

21. During the postnatal period:
    A. The head becomes proportionately smaller
    B. Thoracic and abdominal contours change from round to elliptical
    C. The legs become proportionately longer
    D. The trunk becomes proportionately shorter
    E. All of the above take place during the postnatal period

22. The period of infancy starts at birth and lasts about:
    A. 4 weeks
    B. 4 months
    C. 10 weeks
    D. 12 months
    E. 18 months

23. The lumbar curvature of the spine appears _____ months after birth.
    A. 1-10
    B. 5-8
    C. 8-12
    D. 11-15
    E. 12-18

24. During the first 4 months the birth weight will:
    A. Double
    B. Triple
    C. Quadruple
    D. None of the above is correct

25. At the end of the first year the weight of the baby will have:
    A. Doubled
    B. Tripled
    C. Quadrupled
    D. None of the above is correct

26. The infant is capable of following a moving object with its eyes at:
    A. 2 days
    B. 2 weeks
    C. 2 months
    D. 4 months
    E. 10 months

27. The infant can lift its head and raise its chest at:
    A. 2 months
    B. 3 months
    C. 4 months
    D. 10 months

28. The infant can crawl at:
    A. 2 months
    B. 3 months
    C. 4 months
    D. 10 months
    E. 12 months

29. The infant can stand alone at:
    A. 2 months
    B. 3 months
    C. 4 months
    D. 10 months
    E. 12 months

30. The permanent teeth, with the exception of the third molar, have all erupted by age _____ years.
    A. 6
    B. 8
    C. 12
    D. 14
    E. None of the above is correct

31. Puberty starts at age _____ years in boys:
    A. 10-13
    B. 12-14
    C. 14-16
    D. None of the above is correct

32. Most girls begin breast development at about age:
    A. 8
    B. 9
    C. 10
    D. 11
    E. 12

33. The growth spurt is generally complete by age _____ in males:
    A. 14
    B. 15
    C. 16
    D. 18

34. An average age at which girls begin to menstruate is _____ years.
    A. 10-12
    B. 11-12
    C. 12-13
    D. 13-14
    E. 14-15

35. The first sign of puberty in boys is:
    A. Facial hair
    B. Increased muscle mass
    C. Pubic hair
    D. Deepening of the voice
    E. Increased testicular enlargement

Write the letter of the correct word in the blank next to the appropriate definition.

a. Neonatology   f. Postnatal
b. Neonatal   g. Infancy
c. Adolescence   h. Childhood
d. Deciduous   i. Senescence
e. Puberty

_____ 36. Begins at birth and lasts until death
_____ 37. Concerned with the diagnosis and treatment of disorders of the newborn
_____ 38. Teenage years
_____ 39. From the end of infancy to puberty
_____ 40. Baby teeth
_____ 41. First 4 weeks of infancy
_____ 42. Secondary sexual characteristics occur
_____ 43. Begins at birth and lasts about 18 months
_____ 44. Old age

▷ *If you have had difficulty with this section, review pages 416-420.*

EFFECTS OF AGING

Fill in the blanks.

45. Old bones develop indistinct and shaggy margins with spurs—a process called _____.
46. A degenerative joint disease common in the aged is _____.
47. The number of _____ units in the kidney decreases by almost 50% between the ages of 30 and 75.
48. In old age, respiratory efficiency decreases, and a condition known as _____ _____ results.
49. Fatty deposits accumulate in blood vessels as we age, and the result is _____, which narrows the passageway for the flow of blood.
50. Hardening of the arteries or _____ occurs during the aging process.
51. Another term for high blood pressure is _____.
52. Hardening of the lens is _____.
53. If the lens becomes cloudy and impairs vision, it is called a _____.
54. _____ causes an increase in the pressure within the eyeball and may result in blindness.

▷ *If you have had difficulty with this section, review pages 420-422.*

Unscramble the words

55. ANNFCYI

56. NAALTTSOP

57. OGSSNEGRAONEI

58. GTEYZO

59. HDOOLHCID

Take the circled letters, unscramble them, and fill in the statement.

The secret to Farmer Brown's prize pumpkin crop.

60.

APPLYING WHAT YOU KNOW

61. Heather's mother told the pediatrician during her 1-year visit that she had tripled her birth weight, was crawling actively, and could stand alone. Is Heather's development normal, retarded, or advanced?

62. Clarke is 70 years old. She has always enjoyed food and has had a hearty appetite. Lately, however, she has complained that food "just doesn't taste as good anymore." What might be a possible explanation?

63. Mr. Martinez, age 68, has noticed hearing problems, but only under certain circumstances. He has difficulty with certain tones, especially high or low tones, but has no problem with everyday conversation. What might be a possible explanation?

64. WORD FIND

Can you find 14 terms from the chapter in the box of letters? Words may be spelled top to bottom, bottom to top, right to left, left to right, or diagonally.

```
F  T  P  N  O  I  T  A  T  S  E  G  K  F  U
Z  E  P  O  C  S  O  R  A  P  A  L  H  E  A
E  O  R  I  L  T  O  H  H  O  C  J  V  J  T
C  M  V  T  N  T  I  M  L  N  L  Z  P  U  O
M  A  B  I  I  F  R  M  R  E  D  O  T  C  E
H  Q  C  R  D  L  A  M  E  S  O  D  E  R  M
N  N  M  U  Y  U  I  N  M  F  O  W  G  O  P
K  N  V  T  C  O  C  Z  H  H  F  R  C  N
H  G  Y  R  K  L  T  A  Y  D  U  C  V  P
L  A  T  A  N  T  S  O  P  T  L  U  N  Q  G
Q  N  K  P  Y  Y  S  W  G  A  I  K  E  M  T
U  G  P  Q  N  H  O  Y  X  Y  H  O  Z  B  N
I  Y  T  R  E  B  U  P  L  A  C  E  N  T  A
```

Childhood	Infancy	Parturition
Ectoderm	Laparoscope	Placenta
Embryology	Mesoderm	Postnatal
Fertilization	Morula	Puberty
Gestation	Oviduct	

DID YOU KNOW?

Brain cells do not regenerate. One beer permanently destroys 10,000 cells.

GROWTH/DEVELOPMENT

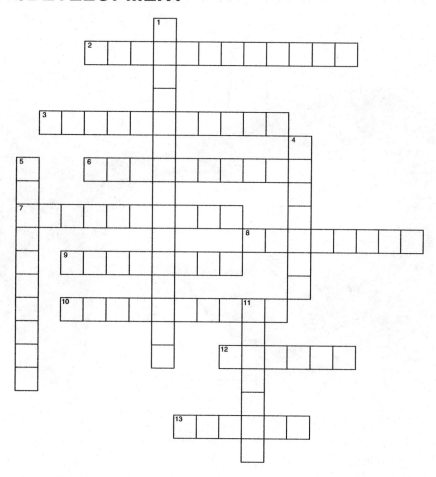

ACROSS

2. Study of how germ layers develop into tissues
3. Process of birth
6. Name of zygote after implantation
7. Science of the development of the individual before birth
8. Eye disease marked by increased pressure in the eyeball
9. Cloudy lens
10. Old age
12. Name of zygote after 3 days
13. Fertilized ovum

DOWN

1. Fatty deposit buildup on walls of arteries
4. First 4 weeks of infancy
5. Hardening of the lens
11. Will develop into a fetal membrane in the placenta

FERTILIZATION AND IMPLANTATION

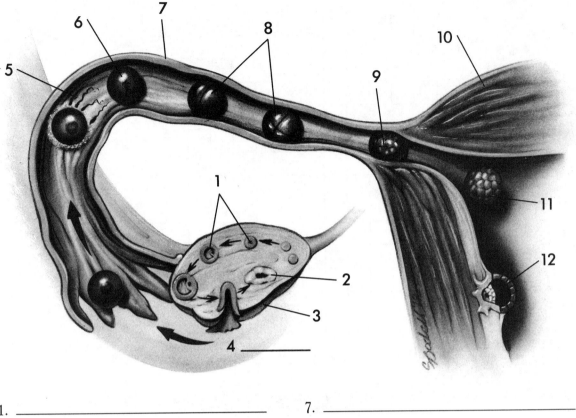

1. _____   7. _____

2. _____   8. _____

3. _____   9. _____

4. _____   10. _____

5. _____   11. _____

6. _____   12. _____

ANSWER KEY

ANSWERS TO EXERCISES

CHAPTER 1
AN INTRODUCTION TO THE STRUCTURE AND FUNCTION OF THE BODY

Matching

1. D, p. 1
2. E, p. 1
3. A, p. 1
4. C, p. 1
5. B, p. 1

Matching

6. C, p. 3
7. A, p. 3
8. E, p. 3
9. D, p. 3
10. B, p. 3

CROSSWORD

11. Superior
12. Inferior
13. Transverse
14. Ventral
15. Lateral
16. Medial
17. Distal

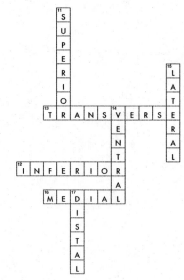

Did you notice that the answers were arranged as they appear on the human body?

Circle the correct answer

18. Inferior, p. 3
19. Anterior, p. 3
20. Lateral, p. 3
21. Proximal, p. 3
22. Superficial, p. 3
23. Equal, p. 4
24. Anterior and posterior, p. 4
25. Upper and lower, p. 4
26. Frontal, p. 4

Select the correct term

27. A, p. 5
28. B, p. 5
29. A, p. 5

30. A, p. 5
31. A, p. 5
32. B, p. 5
33. A, p. 5

Circle the one that does not belong

34. Extremities (all others are part of the axial portions)
35. Cephalic (all others are part of the arm)
36. Plantar (all others are part of the face)
37. Carpal (all others are part of the leg or foot)
38. Tarsal (all others are part of the skull)

Fill in the blanks

39. Survival, p. 8
40. Internal environment, p. 8
41. Feedback loop, p. 10
42. Negative, positive, p. 12
43. Stabilize, p. 12
44. Stimulatory, p. 12
45. Developmental processes, p. 12
46. Aging processes, p. 12

APPLYING WHAT YOU KNOW

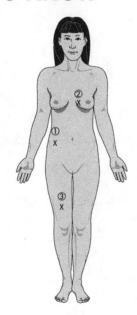

N	H	L	T	E	B	W	N	G	N	M	M	Y	X	A
O	O	H	A	V	U	U	C	L	W	E	P	N	G	L
I	M	U	N	I	T	S	A	I	D	E	M	W	A	L
T	E	R	T	V	C	T	S	I	C	J	S	R	T	K
A	O	N	O	P	T	I	A	I	W	A	T	Y	R	H
Z	S	Q	S	I	R	L	F	A	T	N	R	G	O	W
I	T	W	G	P	R	O	I	R	E	T	S	O	P	F
N	A	A	Z	J	E	E	X	V	E	E	H	L	H	Z
A	S	N	L	M	C	T	P	I	C	P	D	O	Y	T
G	I	C	A	Y	U	O	X	U	M	N	U	I	V	P
R	S	M	R	T	N	U	K	B	S	A	Y	S	V	M
O	R	C	U	R	O	I	R	B	S	Q	L	Y	U	G
L	H	X	E	M	P	M	E	T	S	Y	S	H	V	S
U	W	L	L	Q	D	U	Y	Y	N	E	E	P	J	B
Q	N	Z	P	K	D	B	O	D	C	G	I	N	J	A

Dorsal and Ventral Body Cavities

1. Cranial cavity
2. Spinal cavity
3. Thoracic cavity
4. Mediastinum
5. Abdominal cavity
6. Pelvic cavity

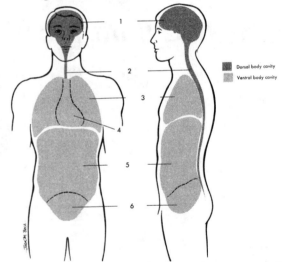

Dorsal body cavity
Ventral body cavity

Directions and Planes of Body

1. Superior
2. Proximal
3. Posterior (dorsal)
4. Anterior (ventral)
5. Inferior
6. Sagittal plane
7. Frontal plane
8. Lateral

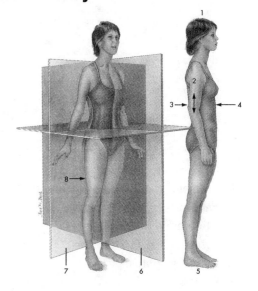

Regions of the Abdomen

1. Epigastric region
2. Left hypochondriac region
3. Umbilical region
4. Left lumbar region
5. Left iliac (inguinal) region
6. Hypogastric region
7. Right iliac (inguinal) region
8. Right lumbar region
9. Right hypochondriac region

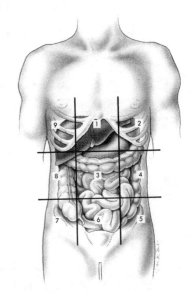

CHAPTER 2
CELLS AND TISSUES

Matching

Group A
1. C, p. 17
2. E, p. 17
3. A, p. 17
4. B, p. 21
5. D, p. 21

Group B
6. D, p. 20
7. E, p. 20
8. A, p. 21
9. B, p. 21
10. C, p. 21

Fill in the blanks

11. Organelles, p.18
12. Tissue typing, p.18
13. Cilia, p.21
14. Aerobic or cellular respiration, p.21
15. Ribosomes, p.20
16. Mitochondria, p.21
17. Lysosomes, p.21
18. Golgi apparatus, p. 21
19. Centrioles, p.21
20. Chromatin granules, p.22

Multiple choice

21. A, p. 22
22. D, p. 23
23. B, p. 23
24. D, p. 25
25. C, p. 25
26. A, p. 25
27. B, p. 25
28. C, p. 25
29. D, p. 25-26
30. A, p. 26
31. B, p. 24
32. A, p. 24

Circle the one that does not belong

33. Uracil (RNA contains the base uracil, not DNA)
34. RNA (the others are complementary base pairings of DNA)
35. Anaphase (the others refer to genes and heredity)
36. Thymine (the others refer to RNA)
37. Interphase (the others refer to translation)
38. Prophase (the others refer to anaphase)
39. Prophase (the others refer to interphase)
40. Metaphase (the others refer to telophase)
41. Gene (the others refer to stages of cell division)

42. Fill in the missing area

TISSUE	LOCATION	FUNCTION
1.	1a.	1a. Absorption by diffusion of respiratory gases between alveolar air and blood

1b.

1b. Absorption by diffusion, filtration and osmosis

2.

2a. Surface of lining of mouth and esophagus

2b. Surface of skin

2.

3.

3. Surface layer of lining of stomach, intestines, and parts of respiratory tract

3.

4. Stratified transitional

4.

4.

5.

5. Surface of lining of trachea

5.

6.

6.

6. Secretion; absorption

1.

1. Between other tissues and organs

1.

2. Adipose

2.

2.

3.

3.

3. Flexible but strong connection

4.

4. Skeleton

4.

5.

5. Part of nasal septum, larynx, rings in trachea and bronchi, disks between vertebrae, external ear

5.

6.

6.

6. Transportation

7. Hemopoietic tissue

7.

7.

1.

1. Muscles that attach to bones, eyeball muscles, upper third of esophagus

1.

2. Cardiac

2.

2.

3.

3. Walls of digestive, respiratory, and genitourinary tracts; walls of blood and large lymphatic vessels; ducts of glands; intrinsic eye muscles; arrector muscles of hair

3.

1.

1. Brain and spinal cord, nerves

1.

APPLYING WHAT YOU KNOW

43.

44. Diffusion
45. Absorption of oxygen into Ms. Bence's blood.
46. Merrily may have exceeded the 20%-22% desirable body fat composition. Fitness depends more on the percentage and ratio of specific tissue types than the overall amount of tissue present.

47. WORD FIND.

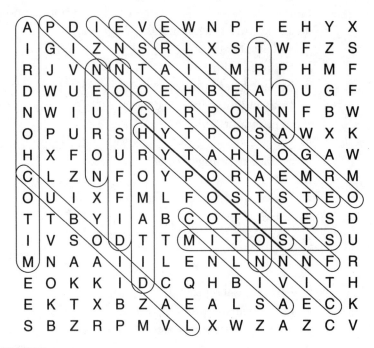

CROSSWORD

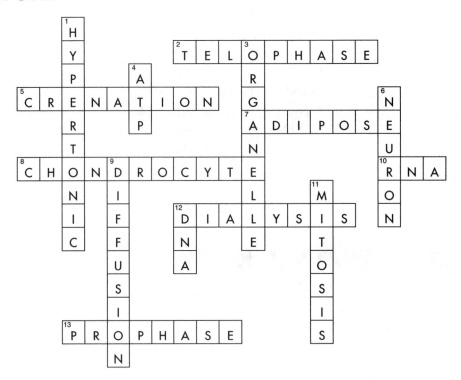

Cell Structure

1. Smooth endoplasmic reticulum
2. Golgi apparatus
3. Nucleolus
4. Nucleus
5. Nuclear membrane
6. Rough endoplasmic reticulum
7. Lysosome
8. Mitochondrion
9. Plasma membrane
10. Smooth endoplasmic reticulum
11. Centrioles
12. Ribosomes

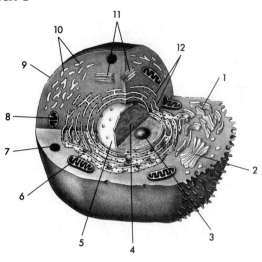

Mitosis

1. Interphase
2. Prophase
3. Metaphase
4. Anaphase
5. Telophase
6. Daughter cells (interphase)

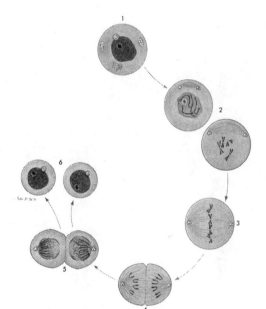

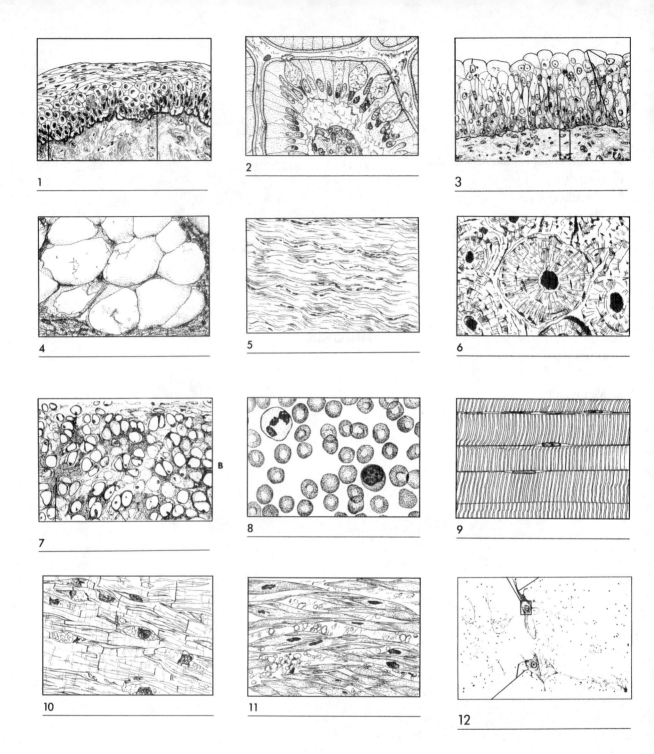

Tissues

1. Stratified squamous epithelium
2. Simple columnar epithelium
3. Stratified transitional epithelium
4. Adipose tissue
5. Dense fibrous connective tissue
6. Bone tissue
7. Cartilage
8. Blood

9. Skeletal muscle
10. Cardiac muscle
11. Smooth muscle
12. Nervous tissue

CHAPTER 3
ORGAN SYSTEMS OF THE BODY
Matching

Group A
1. A, p. 48
2. E, p. 48
3. D, p. 50
4. B, p. 48
5. C, p. 51

Group B
6. F, p. 52
7. E, p. 52
8. B, p. 54
9. A, p. 54
10. C, p. 54
11. D, p. 48

Circle the one that does not belong

12. Mouth (the others refer to the respiratory system)
13. Rectum (the others refer to the reproductive system)
14. Pancreas (the others refer to the circulatory system)
15. Pineal (the others refer to the urinary system)
16. Joints (the others refer to the muscular system)
17. Pituitary (the others refer to the nervous system)
18. Tendons (the others refer to the skeletal system)
19. Appendix (the others refer to the endocrine system)
20. Thymus (the others refer to the integumentary system)
21. Trachea (the others refer to the digestive system)
22. Liver (the others refer to the lymphatic system)

Fill in the missing area

SYSTEM	ORGAN	FUNCTIONS
23.		Protection, regulation of body temperature, synthesis of chemicals and hormones, serves as a sense organ
24.	Bones, joints	
25.		Movement, maintains body posture, produces heat
26. Nervous		

27.   Pituitary, thymus, pineal, adrenal, hypothalamus, thyroid, pancreas, parathyroid, ovaries, testes

28.   Transportation, immunity

29.   Lymph nodes, lymph vessels, thymus, spleen, tonsils

30. Urinary

31.   Mouth, pharynx, esophagus, stomach, small and large intestine, rectum, anal canal, teeth, salivary glands, tongue, liver, gallbladder, pancreas, appendix

32. Respiratory

33.   a. Gonads—testes and ovaries
               b. Accessory glands (p. 60) Supporting structures (p. 60)

Unscramble the words

34. Heart
35. Pineal
36. Nerve
37. Esophagus
38. Nervous

APPLYING WHAT YOU KNOW

39. (a) Endocrinology (endocrine system)
     (b) Gynecology (reproductive system)

40. The skin protects the underlying tissue against invasion by harmful bacteria. With a large percentage of Brian's skin destroyed, he was vulnerable to bacteria, and so he was placed in the cleanest environment possible—isolation. Jenny is required to wear special attire so that the risk of a visitor bringing bacteria to the patient is reduced.

41. WORD FIND

CROSSWORD

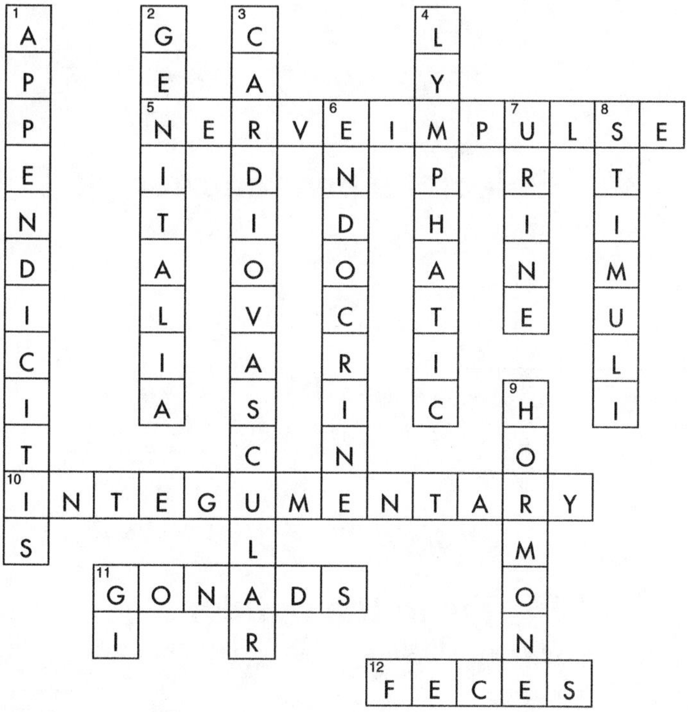
1 APPENDICITIS
2 GENITALIA
3 CARDIOVASCULAR
4 LYMPHATIC
5 NERVE IMPULSE
6 ENDOCRINE
7 URINE
8 STIMULI
9 HORMONE
10 INTEGUMENTARY
11 GONADS
12 FECES
GI

CHAPTER 4
THE INTEGUMENTARY SYSTEM AND BODY MEMBRANES

Select the best answer

1. B, p. 63
2. D, p. 65
3. C, p. 65
4. A, p. 63
5. B, p. 63
6. D, p. 65
7. C, p. 65
8. C, p. 65

Matching

Group A

9. D, p. 63
10. A, p. 65
11. B, p. 65
12. C, p. 68
13. E, p. 65

Group B
14. A, p. 67
15. D, p. 67
16. E, p. 67
17. C, p. 68
18. B, p. 67

Select the best answer

19. A, p. 66
20. B, p. 68
21. B, p. 68
22. A, p. 68
23. A, p. 67
24. B, p. 68
25. B, p. 68
26. B, p. 68
27. B, p. 68
28. A, p. 67 (text figure 4-3)

Fill in the blanks

29. Protection, temperature regulation, and sense organ activity, p. 73
30. Melanin, p. 73
31. Lanugo, p. 68
32. Hair papillae, p. 69
33. Protein, p. 71
34. Arrector pili, p. 70
35. Light touch, p. 71
36. Eccrine, p. 72
37. Apocrine, p. 72
38. Sebum, p. 72

Circle the correct answer

39. Will not, p. 75
40. Will, p. 75
41. Will not, p. 75
42. 11, p. 74
43. Third, p. 75

Unscramble the words

44. Epidermis
45. Keratin
46 Hair
47. Lanugo
48. Dehydration
49. Third Degree

APPLYING WHAT YOU KNOW

50. 46%
51. Pleurisy
52. Fingerprints

53. WORD FIND

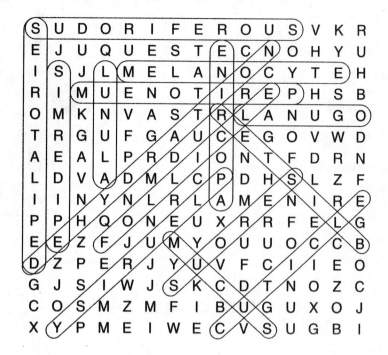

CROSSWORD

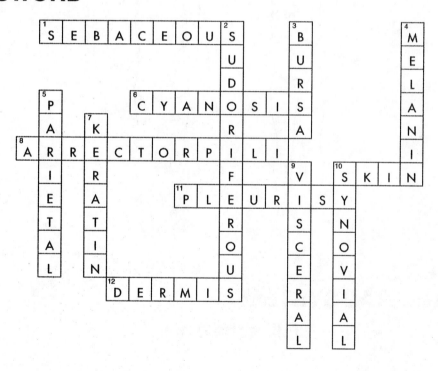

Longitudinal Section of the Skin

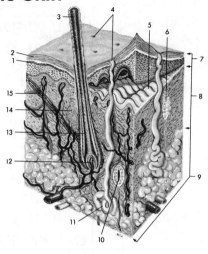

1. Pigment layer
2. Stratum corneum
3. Hair shaft
4. Openings of sweat ducts
5. Dermal papilla
6. Meissner's corpuscle
7. Epidermis
8. Dermis
9. Subcutaneous fatty tissue
10. Pacinian corpuscle
11. Sweat gland
12. Papilla of hair
13. Hair follicle
14. Sebaceous (oil) gland
15. Arrector pili muscle

Rule of Nine

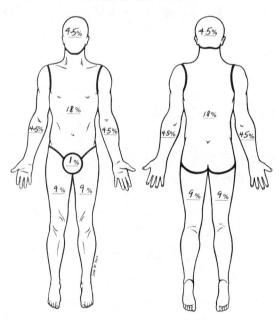

CHAPTER 5
THE SKELETAL SYSTEM

Fill in the blanks

1. 4, p. 81
2. Medullary cavity, p. 82
3. Articular cartilage p. 82
4. Endosteum, p. 82
5. Hemopoiesis, p. 81
6. Red bone marrow, p. 81
7. Periosteum, p. 82
8. Elderly white female, p. 86

9. Calcium, p. 81
10. Move, p. 81

Matching

Group A
11. D, p. 82
12. B, p. 82
13. E, p. 82
14. A, p. 82
15. C, p. 84

Group B
16. D, p. 82
17. A, p. 84
18. E, p. 82
19. B. p. 84
20. C, p. 84

True or false

21. T, p. 84
22. Epiphyses, not diaphyses, p. 86
23. Osteoblasts, not osteoclasts, p. 84
24. T, p. 84
25. Increase, not decrease, p. 84
26. Juvenile, not adult, p. 86
27. Diaphysis, not articulation, p. 86
28. T, p. 84
29. Ceases, not begins, p. 86
30. T, p. 86

Multiple choice

31. A, p. 87
32. D, p. 90
33. A, p. 94
34. D, p. 94
35. C, p. 95
36. C, p. 99
37. D, p. 95
38. D, p. 99
39. A, p. 99
40. B, p. 90
41. A, p. 92
42. B, p. 97
43. B, p. 94
44. B, p. 95
45. C, p. 99
46. A, p. 99
47. D, p. 90

48. C, p. 92
49. C, p. 90

Circle the one that does not belong

50. Coxal bone (all others refer to the spine)
51. Axial (all others refer to the appendicular skeleton)
52. Maxilla, (all others refer to the cranial bones)
53. Ribs (all others refer to the shoulder girdle)
54. Vomer (all others refer to the bones of the middle ear)
55. Ulna (all others refer to the coxal bone)
56. Ethmoid (all others refer to the hand and wrist)
57. Nasal (all others refer to cranial bones)
58. Anvil (all others refer to the cervical vertebra)

Choose the correct answer

59. A, p. 100
60. B, p. 100
61. B, p. 86
62. A, p. 100
63. B, p. 100

Matching

64. C, p. 90
65. G, p. 94
66. J, L, M and K, p. 100
67. N, p. 98
68. I, p. 97
69. A, p. 90
70. P, p. 99
71. D, B, p. 90
72. F, p. 90
73. H, Q, p. 97
74. O, T, p. 100
75. R, p. 90
76. S, E, p. 90

Choose the correct answer

77. Diarthroses, p. 102
78. Synarthrotic, p. 102
79. Diarthrotic, p. 103
80. Ligaments, p. 104
81. Articular cartilage, p. 104
82. Least movable, p. 105
83. Largest, p. 106
84. 2, p. 105
85. Mobility, p. 105
86. Pivot, p. 105

Unscramble the words

87. Vertebrae
88. Pubis
89. Scapula
90. Mandible
91. Phalanges
92. Pelvic girdle

APPLYING WHAT YOU KNOW

93. The bones are responsible for the majority of our blood cell formation. The disease condition of the bones might be inhibiting the production of blood cells for Mrs. Perine.
94. Epiphyseal cartilage is present only while a child is still growing. It becomes bone in adulthood. It is particularly vulnerable to fractures in childhood and preadolescence.
95. Osteoporosis
96. WORD FIND

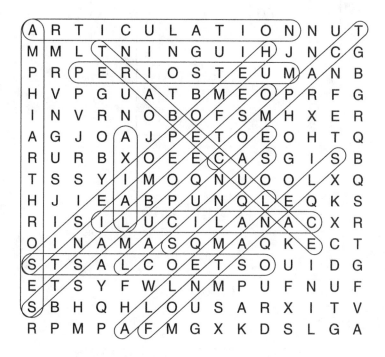

CROSSWORD

Long Bone

1. Articular cartilage
2. Spongy bone
3. Epiphyseal plate
4. Red marrow cavities
5. Compact bone
6. Medullary cavity
7. Yellow marrow
8. Periosteum

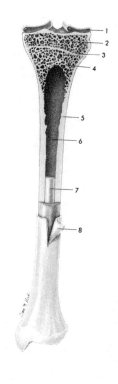

Anterior View of Skeleton

1. Orbit
2. Mandible
3. Sternum
4. Xiphoid process
5. Costal cartilage
6. Coxal bone
7. Ilium
8. Pubis
9. Ischium
10. Frontal
11. Nasal
12. Maxilla
13. Clavicle
14. Ribs
15. Humerus
16. Vertebral column
17. Ulna
18. Radius
19. Sacrum
20. Coccyx
21. Carpals
22. Metacarpals
23. Phalanges
24. Femur
25. Patella
26. Tibia
27. Fibula
28. Tarsals
29. Metatarsals
30. Phalanges

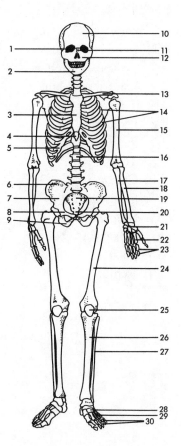

Posterior View of Skeleton

1. Parietal
2. Cervical vertebrae
3. Thoracic vertebrae
4. Lumbar vertebrae
5. Coccyx
6. Femur
7. Fibula
8. Tibia
9. Calcaneus
10. Occipital
11. Scapula
12. Coxal (hip) bone
13. Sacrum
14. Ischium

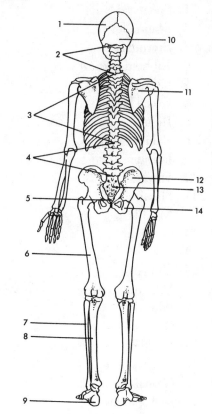

Skull, Right Side

1. Parietal bone
2. Squamous suture
3. Occipital bone
4. Lambdoidal suture
5. Temporal bone
6. External auditory canal
7. Mastoid process
8. Coronal suture
9. Frontal bone
10. Sphenoid bone
11. Ethmoid bone
12. Nasal bone
13. Zygomatic bone
14. Maxilla
15. Mandible

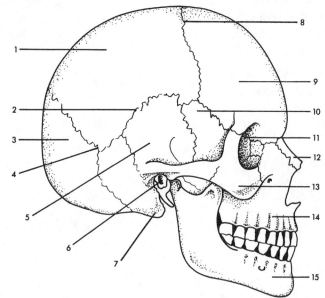

Skull, Front view

1. Sphenoid bone
2. Ethmoid bone
3. Lacrimal bone
4. Zygomatic bone
5. Vomer
6. Frontal bone
7. Parietal bone
8. Nasal bone
9. Inferior concha
10. Maxilla
11. Mandible

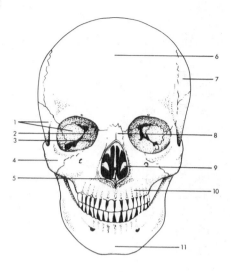

Structure of a Diarthrotic Joint

1. Bone
2. Synovial membrane
3. Blood vessel
4. Nerve
5. Joint capsule
6. Periosteum
7. Bone
8. Articular cartilage
9. Bursa
10. Joint capsule

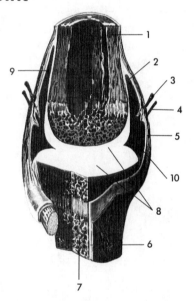

CHAPTER 6
THE MUSCULAR SYSTEM

Select the correct term

1. A, p. 111
2. B, p. 111
3. C, p. 111
4. C, p. 111
5. A, p. 111
6. B, p. 111
7. C and B, p. 111
8. A, p. 111
9. C, p. 111
10. C, p. 111

Matching

Group A

11. D, p. 111
12. B, p. 111
13. A, p. 111
14. E, p. 111
15. C, p. 112

Group B

16. E, p. 113
17. C, p. 113
18. B, p. 113
19. A, p. 113
20. D, p. 113

Fill in the blanks

21. Pulling, p. 113
22. Insertion, p. 113
23. Insertion, origin, p. 113
24. Prime mover, p. 113
25. Antagonist, p. 113
26. Synergist, p. 113
27. Tonic contraction, p. 115
28. Muscle tone, p. 115
29. Hypothermia, p. 115
30. ATP, p. 115

True or false

31. Neuromuscular junction, p. 116
32. T, p. 116
33. T, p. 115
34. Oxygen debt, p. 115
35. "All or none," p. 117
36. Lactic acid, p. 115
37. T, p. 117
38. T, p. 117
39. Skeletal muscle, p. 116
40. T, p. 116

Multiple choice

41. A, p. 117
42. B, p. 117
43. B, p. 117
44. C, p. 118
45. D, p. 118
46. A, p. 117
47. B, p. 117
48. C, p. 117
49. B, p. 118
50. D, p. 118

51. C, p. 122
52. F, p. 123; A and D, p. 128
53. F, p. 123; and B, p. 128
54. A, p. 122
55. C, p. 128
56. B, p. 123 ; and F, p. 123
57. A, p. 122
58. A and D, p. 122
59. B, p. 122
60. B, p. 122
61. A, p. 123; E, p. 122
62. B, p. 122
63. D, p. 122

Multiple choice

64. A, p. 127
65. D, p. 127
66. C, p. 127
67. A, p. 127
68. D, p. 127
69. C, p. 127

APPLYING WHAT YOU KNOW

70. Bursitis
71. Deltoid area
72. Tendon
73. WORD FIND

CROSSWORD

```
            1H              2R O T A T I O N              5I
      4P    Y                  B                           N
6A T R O P H Y            7A D D U C T I O N                S
   O     E                   C  U                          E
   N     R        8O         T  C                          R
9A N T A G O N I S T         N  C                          T
   T     R                      R     10S                  I
   I     R        11T         I  G     Y                   O
   O     O        12E X T E N S I O N  N                   N
   N     P        N             N      E
         H        D                    R
         Y        O                    G
                  N                    I
              13B U R S A              T
```

Muscles, Anterior View

1. Sternocleidomastoid
2. Trapezius
3. Pectoralis major
4. Rectus abdominis
5. External abdominal oblique
6. Iliopsoas
7. Quadriceps group
8. Tibialis anterior
9. Peroneus longus
10. Peroneus brevis
11. Soleus
12. Gastrocnemius
13. Sartorius
14. Adductor group
15. Brachialis
16. Biceps brachii
17. Deltoid
18. Facial muscles

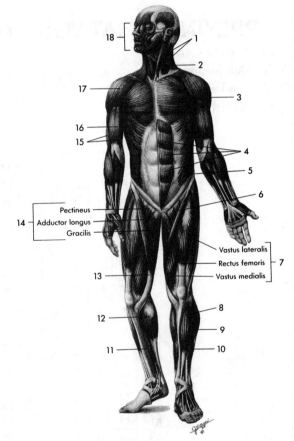

Muscles, Posterior View

1. Trapezius
2. External abdominal oblique
3. Gluteus maximus
4. Adductor magnus
5. Soleus
6. Peroneus brevis
7. Peroneus longus
8. Gastrocnemius
9. Hamstring group
10. Latissimus dorsi
11. Triceps brachii
12. Deltoid
13. Sternocleidomastoid

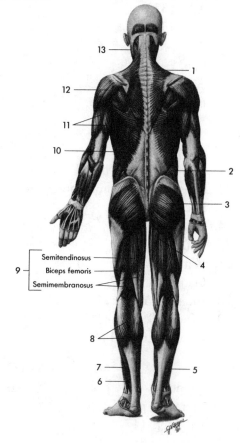

CHAPTER 7
THE NERVOUS SYSTEM

Matching

Group A
1. B. p. 137
2. C, p. 137
3. D, p. 137
4. A, p. 137

Group B
5. B, p. 137
6. D, p. 138
7. C, p. 137
8. A, p. 138
9. F, p. 140
10. E, p. 141

Select the best choice

11. A, p. 137
12. B, p. 138

13. B, p. 140
14. A, p. 137
15. A, p. 137
16. B, p. 140
17. B, p. 140
18. A, p. 137
19. B, p. 143
20. A, p. 138

Fill in the blanks

21. Two-neuron arc, p. 141
22. Sensory, interneurons, and motor neurons, p. 142
23. Receptors, p. 142
24. Synapse, p. 142
25. Reflex, p. 142
26. Withdrawal reflex, p. 143
27. Ganglion, p. 142
28. Interneurons, p. 143
29. "Knee jerk," p. 142
30. Gray matter, p. 143

Circle the correct word

31. Do not, p. 143
32. Increases, p. 145
33. Excess, p. 145
34. Postsynaptic, p. 145
35. Presynaptic, p. 145
36. Neurotransmitter, p. 146
37. Communicate, p. 146
38. Specifically, p. 146
39. Sleep, p. 146
40. Pain, p. 146

Multiple choice

41. E, p. 147
42. D, p. 147
43. A, p. 147
44. E, p. 149
45. E, p. 147-149
46. D, p. 149
47. B, p. 150
48. E, p. 150
49. B, p. 150
50. D, p. 150
51. D, p. 149
52. B, p. 150
53. A, p. 150
54. D, p. 150
55. C, p. 150

True or false

56. 17 to 18 inches, p. 150
57. Bottom of the first lumbar vertebra, p. 150
58. Lumbar punctures, not CAT scan, p. 156
59. Spinal tracts, not dendrites, p. 152
60. T, p. 152
61. One general function, not several, p. 152
62. Anesthesia, not paralysis, p. 154

Circle the one that does not belong

63. Ventricles (all others refer to meninges)
64. CSF (all others refer to the arachnoid)
65. Pia mater (all others refer to the cerebrospinal fluid)
66. Choroid plexus (all others refer to the dura mater)
67. Brain tumor (all others refer to a lumbar puncture)

CRANIAL NERVES

68.

NERVE	CONDUCT IMPULSES	FUNCTION
I Olfactory		
II		Vision
III	From brain to eye muscles	
IV Trochlear		
V		Sensations of face, scalp, and teeth, chewing movements
VI	From brain to external eye muscles	
VII		Sense of taste; contractions of muscles of facial expression
VIII Vestibulocochlear		
IX	From throat and taste buds of tongue to brain; also from brain to throat muscles and salivary glands	
X Vagus		
XI		Shoulder movements; turning movements of head
XII Hypoglossal		

Select the best choice

69. A, p. 156
70. B, p. 159
71. A, p. 156
72. B, p. 166
73. B, p. 156
74. A, p. 156
75. B, p. 158
76. B, p. 159

Matching

77. D, p. 160
78. E, p. 160
79. F, p. 160
80. B, p. 162
81. A, p. 160
82. C, p. 160

Multiple choice

83. C, p. 163
84. B, p. 163
85. B, p. 163
86. D, p. 163
87. A, p. 163
88. A, p. 163

Choose the correct response

89. B, p. 164
90. A, p. 164
91. A, p. 164
92. B, p. 164
93. A, p. 164
94. B, p. 164
95. A, p. 164
96. A, p. 164
97. B, p. 164
98. B, p. 164

Fill in the blanks

99. Acetylcholine, p. 164
100. Adrenergic fibers, p. 164
101. Cholinergic fibers, p. 164
102. Homeostasis, p. 164
103. Heart rate, p. 165
104. Decreased, p. 165

Unscramble the words

105. Neurons
106. Synapse
107. Autonomic
108. Smooth muscle
109. Sympathetic

APPLYING WHAT YOU KNOW

110. Right
111. Hydrocephalus
112. Sympathetic
113. Parasympathetic
114. Sympathetic; No, the digestive process is not active during sympathetic control. Bill may experience nausea, vomiting, or discomfort because of this factor.
115. WORD FIND

CROSSWORD

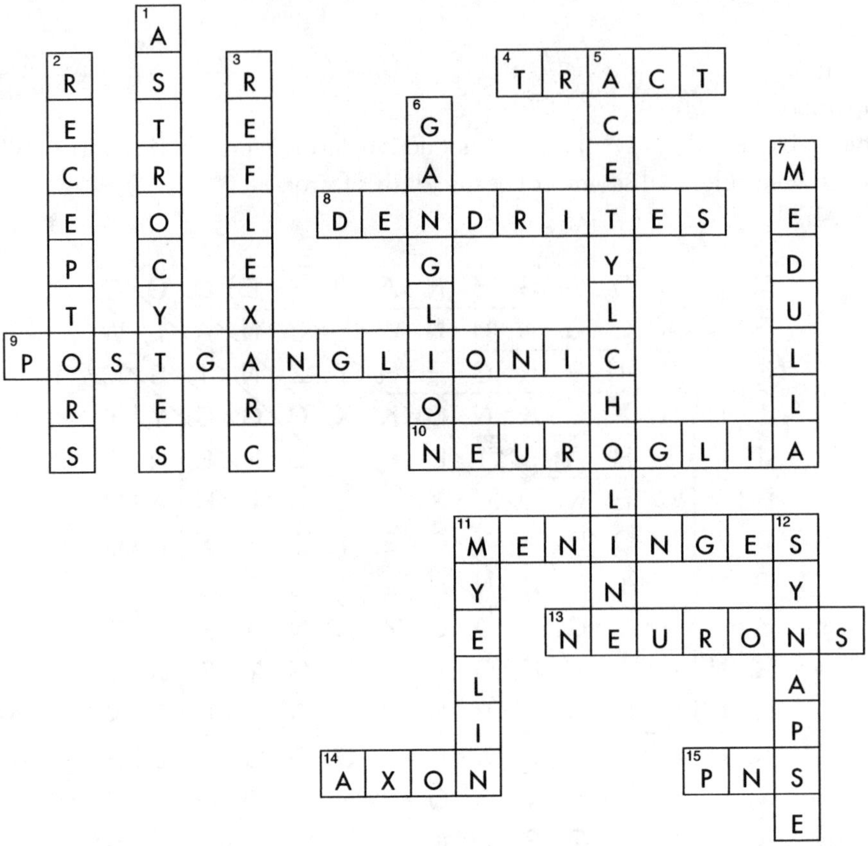

1 ASTROCYTES
2 RECEPTORS
3 REFLEXARC
4 TRACT
5 ACETYLCHOLINE
6 GANGLIO
7 MEDULLA
8 DENDRITES
9 POSTGANGLIONIC
10 NEUROGLIA
11 MYELIN
12 SYNAPSE
13 NEURONS
14 AXON
15 PNS
MENINGES

Neuron

1. Dendrites
2. Cell body
3. Nucleus
4. Axon
5. Schwann cell
6. Myelin
7. Mitochondrion

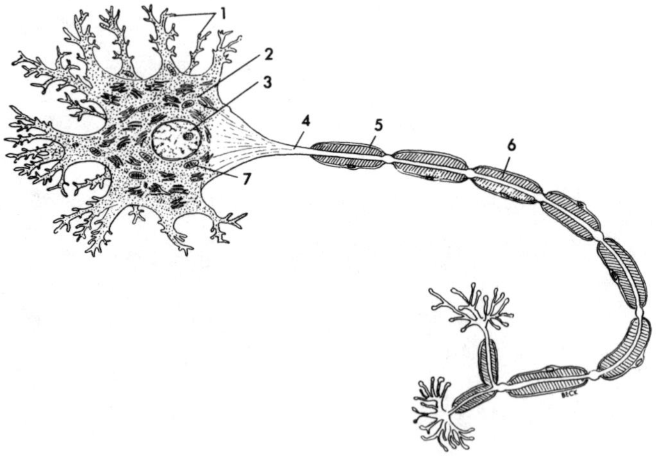

1
2
3
4
5
6
7

1. Olfactory nerve
2. Trigeminal nerve
3. Glossopharyngeal nerve
4. Hypoglossal nerve
5. Accessory nerve
6. Vagus nerve
7. Vestibulocochlear nerve
8. Facial nerve
9. Abducens nerve
10. Oculomoter nerve
11. Optic nerve
12. Trochlear nerve

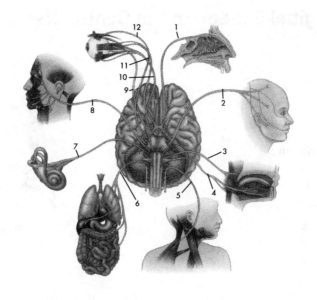

Patellar Reflex

1. Dorsal root ganglion
2. Sensory neuron
3. Stretch receptor
4. Patella
5. Patellar tendon
6. Quadriceps muscle
7. Motor neuron
8. Monosynaptic synapse
9. Gray matter
10. Interneuron

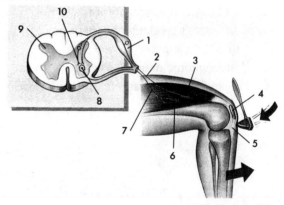

The Cerebrum

1. Occipital lobe
2. Temporal lobe
3. Lateral fissure
4. Frontal lobe
5. Central sulcus
6. Parietal lobe

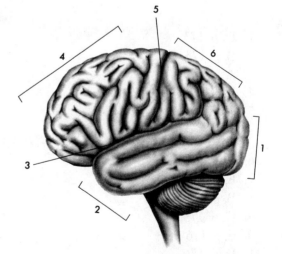

Sagittal Section of the Central Nervous System

1. Skull
2. Pineal gland
3. Cerebellum
4. Midbrain
5. Spinal cord
6. Medulla
7. Reticular formation
8. Pons
9. Pituitary gland
10. Hypothalamus
11. Cerebral cortex
12. Thalamus
13. Corpus callosum

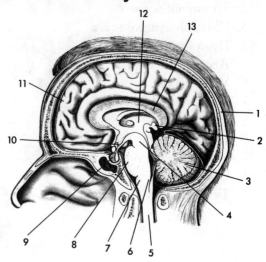

Neuron Pathways

1. Somatic motor neuron's axon
2. Cell body of somatic motor neuron
3. Cell body of preganglionic neuron
4. Collateral ganglion
5. Postganglionic neuron's axon
6. Preganglionic sympathetic neuron's axon

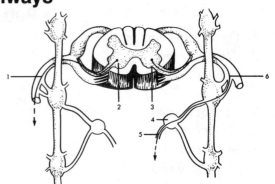

CHAPTER 8
SENSE ORGANS

Matching

1. D, p. 173
2. B, p. 173
3. A, p. 176
4. E, p. 174
5. C, p. 173

Multiple choice

6. C, p. 176
7. E, p. 176
8. B, p. 176
9. C, p. 176
10. E, p. 176
11. D, p. 176
12. A, p. 176
13. B, p. 176

14. C, p. 177
15. B, p. 177
16. D, p. 176
17. A, p. 176
18. D, p. 179

Select the best choice

19. B, p. 181
20. C, p. 182
21. B, p. 181
22. A, p. 181
23. C, p. 182
24. A, p. 180
25. C, p. 182
26. B, p. 181
27. B, p. 182
28. C, p. 183

Fill in the blanks

29. Auricle and external auditory canal, p. 180
30. Eardrum, p. 181
31. Ossicles, p. 181
32. Oval window, p. 181
33. Otitis media, p. 182
34. Vestibule, p. 182
35. Mechanoreceptors, p. 183
36. Crista ampullaris, p. 183

Circle the correct answer

37. Papillae, p. 184
38. Cranial, p. 185
39. Mucus, p. 185
40. Memory, p. 185
41. Chemoreceptors, p. 184,185

APPLYING WHAT YOU KNOW

42. External otitis
43. Cataracts
44. The eustachian tube connects the throat to the middle ear and provides a perfect pathway for the spread of infection.
45. Olfactory

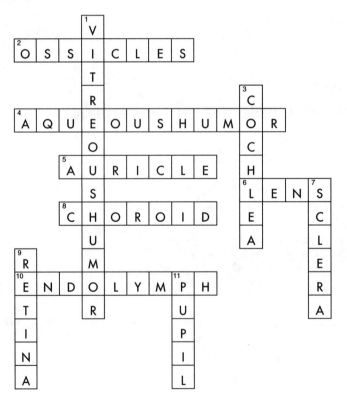

CROSSWORD

Eye

1. Conjunctiva
2. Iris
3. Lens
4. Ciliary muscle
5. Retina
6. Choroid layer
7. Sclera
8. Central retinal artery and vein
9. Optic nerve
10. Optic disc (blind spot)
11. Medial rectus muscle
12. Anterior cavity
13. Pupil
14. Cornea

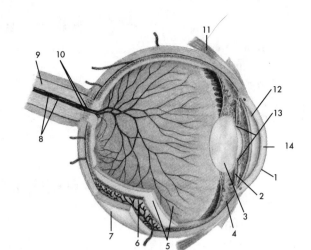

Ear

1. Temporal bone
2. External auditory meatus
3. Auricle (pinna)
4. Tympanic membrane (eardrum)
5. Malleus
6. Incus
7. Stapes
8. Auditory (Eustachian tube)
9. Cochlea
10. Cochlear nerve
11. Semicircular canals

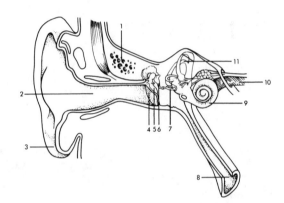

CHAPTER 9
ENDOCRINE SYSTEM

Matching

Group A
1. D, p. 192
2. C, p. 192
3. E, p. 192
4. A, p. 192
5. B, p. 192

Group B
6. E, p. 194
7. C, p. 195
8. A, p. 191

9. D, p. 191
10. B, p. 191

Fill in the blanks

11. Second messenger, p. 191
12. Recognize, p. 191
13. First messengers, p. 191
14. Target organs, p. 191
15. Cyclic AMP, p. 191
16. Target cells, p. 191
17. Steroid abuse, p. 192

Multiple choice

18. B, p. 196
19. E, p. 196
20. D, p. 197
21. D, p. 198
22. A, p. 198
23. C, p. 198
24. C, p. 198
25. B, p. 196
26. A, p. 197
27. A, p. 197
28. D, p. 198
29. B, p. 198
30. A, p. 198
31. C, p. 198
32. C, p. 198

Select the best answer

33. A, p. 196
34. B, p. 196
35. B, p. 198
36. C, p. 199
37. A, p. 197
38. C, p. 199
39. A, p. 198
40. A, p. 196
41. A, p. 197
42. C, p. 199

Circle the correct term

43. Below, p. 200
44. Calcitonin, p. 200
45. Iodine, p. 200
46. Do not, p. 200
47. Thyroid, p. 200,201
48. Decrease, p. 202
49. Hypothyroidism, p. 201
50. Cretinism, p. 201

51. PTH, p. 202
52. Increase, p. 202

Fill in the blanks

53. Adrenal cortex and adrenal medulla, p. 203
54. Corticoids, p. 204
55. Mineralocorticoids, p. 204
56. Glucocorticoids, p. 204
57. Sex hormones, p. 205
58. Gluconeogenesis, p. 204
59. Blood pressure, p. 204
60. Epinephrine and norepinephrine, p. 205
61. Stress, p. 205
62. General adaptation syndrome, p. 206

Select the best response

63. A, p. 206
64. A, p. 204
65. B, p. 205
66. A, p. 206
67. B, p. 206
68. A, p. 204
69. A, p. 205

Circle the term that does not belong

70. Beta cells (all others refer to glucagon)
71. Glucagon (all others refer to insulin)
72. Thymosin (all others refer to female sex glands)
73. Chorion (all other refer to male sex glands)
74. Aldosterone (all others refer to the thymus)
75. ACTH (all others refer to the placenta)
76. Semen (all others refer to the pineal gland)

Matching

Group A
77. E, p. 207
78. C, p. 207
79. B, p. 208
80. D, p. 209
81. A, p. 208

Group B
82. E, p. 209
83. A, p. 209
84. B, p. 209
85. C, p. 209
86. D, p. 209

APPLYING WHAT YOU KNOW

87. She was pregnant
88. Zona reticularis of the adrenal cortex
89. WORD FIND

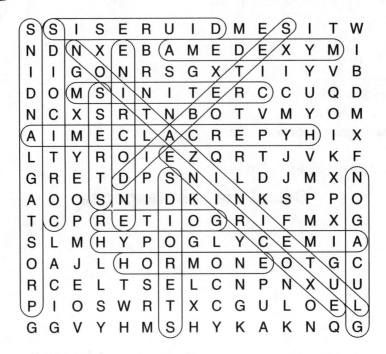

CROSSWORD

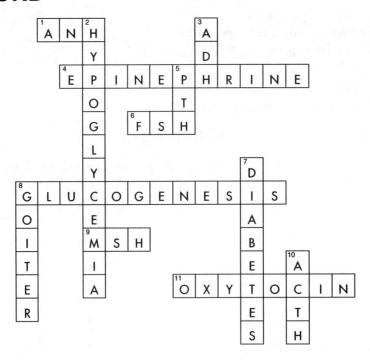

Endocrine Glands

1. Pineal
2. Pituitary
3. Parathyroids
4. Thymus
5. Adrenal
6. Pancreas
7. Ovaries
8. Testes
9. Thyroid

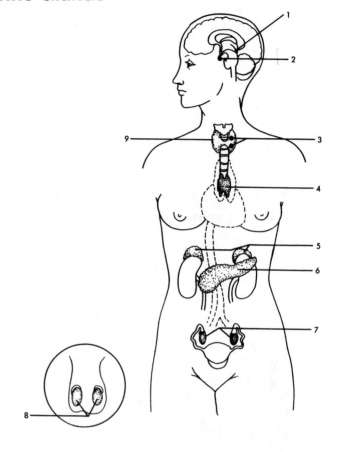

CHAPTER 10
BLOOD

Multiple choice

1. E, p. 215
2. D, p. 215
3. D, p. 215
4. B, p. 215
5. C, p. 215
6. A, p. 218
7. B, p. 218
8. D, p. 215
9. C, p. 220
10. D, p. 218
11. A, p. 218
12. B, p. 223
13. B, p. 221
14. B, p. 215
15. E, p. 217
16. B, p. 216
17. C, p. 216
18. D, p. 216

19. D, p. 218
20. D, p. 218
21. B, p. 220
22. D, p. 215
23. B, p. 219
24. C, p. 219 and p. 220
25. B, p. 219
26. A, p. 220
27. E, p. 221

Fill in the blank areas

28. p. 222-223

Blood Type	Antigen	Antibody
A	A	Anti-B
B	B	Anti-A
AB	A,B	None
O	None	Anti-A, Anti-B

Fill in the blanks

29. Antigen, p. 222
30. Antibody, p. 222
31. Agglutinate, p. 222
32. Erythroblastosis fetalis, p. 223
33. Rhesus monkeys, p. 223
34. RhoGAM, p. 223
35. AB, p. 223

APPLYING WHAT YOU KNOW

36. No. If Mrs. Payne were a negative Rh factor and her husband were a positive Rh factor, it would set up the strong possibility of erythroblastosis fetalis.
37. Both procedures assist the clotting process.

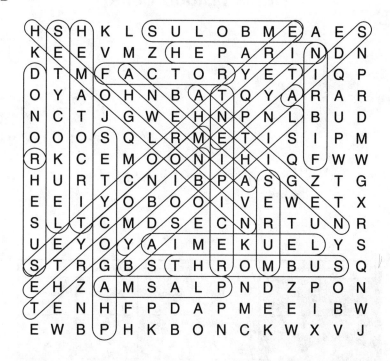

CROSSWORD

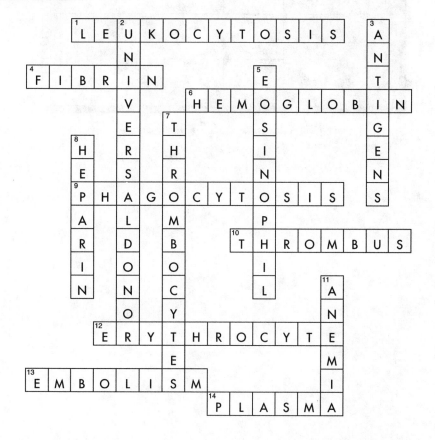

Human Blood Cells

1. Red blood cells
2. Platelets
3. Basophil
4. Neutrophil
5. Eosinophil
6. Lymphocyte
7. Monocyte

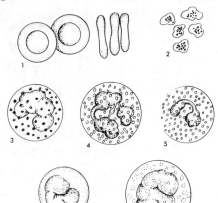

Recipient's blood		Reactions with donor's blood			
RBC antigens	Plasma antibodies	Donor type O	Donor type A	Donor type B	Donor type AB
None (Type O)	Anti-A Anti-B				
A (Type A)	Anti-B				
B (Type B)	Anti-A				
AB (Type AB)	(none)				

 Normal blood    Agglutinated blood

CHAPTER 11
THE CIRCULATORY SYSTEM

Fill in the blanks

1. CPR, p. 229
2. Interatrial septum, p. 232
3. Atria, p. 229
4. Ventricles, p. 229
5. Myocardium, p. 229
6. Endocarditis, p. 232
7. Bicuspid or mitral and tricuspid, p. 232
8. Pulmonary circulation, p. 234
9. Coronary embolism or coronary thrombosis, p. 234
10. Myocardial infarction, p. 234
11. Sinoatrial node, p. 237
12. P, QRS complex, and T, p. 238
13. Repolarization, p. 238

Select the best answer

14. A, p. 232
15. K, p. 229
16. G, p. 234
17. C, p. 234
18. D, p. 232
19. F, p. 229
20. H, p. 234
21. B, p. 234
22. E, p. 238
23. I, p. 235
24. J, p. 238
25. L, p. 229
26. M, p. 232

Matching

27. D, p. 238
28. B, p. 238
29. C, p. 238
30. G, p. 238
31. A, p. 241
32. E, p. 238
33. F, p. 238

Multiple choice

34. D, p. 238
35. C, p. 238
36. B, p. 241
37. A, p. 238
38. B, p. 241
39. B, p. 245

40. D, p. 246
41. B, p. 248
42. B, p. 248
43. A, p. 248
44. D, p. 238
45. A, p. 241

True or false

46. Highest in arteries, lowest in veins, p. 248
47. Blood pressure gradient, p. 248
48. Stop, p. 248
49. High, p. 248
50. Decreases, p. 248
51. T, p. 249
52. T, p. 249
53. T, p. 249
54. Stronger will increase, weaker will decrease, p. 250
55. Contract, p. 251
56. Relax, p. 251
57. Artery, p. 251
58. T, p. 252
59. T, p. 252
60. Brachial, p. 252

Unscramble the words

61. Systemic
62. Venule
63. Artery
64. Pulse
65. Vessel

APPLYING WHAT YOU KNOW

66. Coronary bypass surgery
67. Artificial pacemaker
68. The endocardial lining can become rough and abrasive to red blood cells passing over its surface. As a result, a fatal blood clot may be formed.
69. Mr. Philbrick may be hemorrhaging. The heart beats faster during hemorrhage in an attempt to compensate for blood loss.

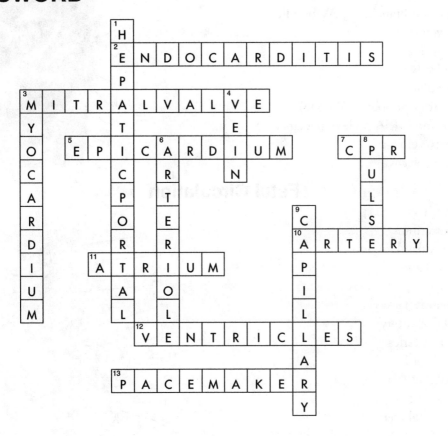

Y H Y S I S O B M O R H T A R
L A C I L I B M U O A Y N I Y
E E S Y S T E M I C C G D E U
M V E N U L E D D C I X M D I
E Y M U I R T A R N L E C G Y
N O I T A Z I R A L O P E D N
U D L A T R O P C I T A P E H
S I U K M U E T O E S L U P U
R P N F Y C B E D R A L Z P J
D S A H T I T V N L I I B D W
Q U R O I V A Q E O D A K R K
K C R M P G A I G N D D Z Y J
Y I Y R I N E A F Z L Q S P X
S R M I C C Q O A W U N H O K
P T A V H H Z L H I J J X K Z

CROSSWORD

(2 across) ENDOCARDITIS
(3 across) MITRALVALVE
(5 across) EPICARDIUM
(7 across) CPR
(10 across) ARTERY
(11 across) ATRIUM
(12 across) VENTRICLES
(13 across) PACEMAKER

1 HEPATIC
3 MYOCARDIUM
4 VEIN
6 ARTERIOL...
8 PULS...
9 CAPILLARY

The Heart

1. Aorta
2. Pulmonary arteries
3. Left pulmonary veins
4. Left atrium
5. Aortic semilunar valve
6. Bicuspid valve
7. Chordae tendineae
8. Left ventricle
9. Interventricular septum
10. Right ventricle
11. Tricuspid valve
12. Right atrium
13. Pulmonary semilunar valve
14. Superior vena cava

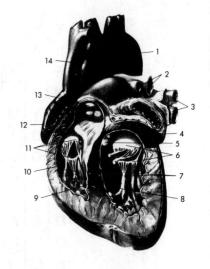

Conduction System of the Heart

1. Aorta
2. Pulmonary artery
3. Pulmonary veins
4. Mitral (bicuspid) valve
5. Left ventricle
6. Right and left branches of AV bundle
7. Inferior vena cava
8. Right ventricle
9. Tricuspid valve
10. Right atrium
11. Atrioventricular node (AV node)
12. Sinoatrial node (SA node or pacemaker)
13. Pulmonary veins
14. Superior vena cava

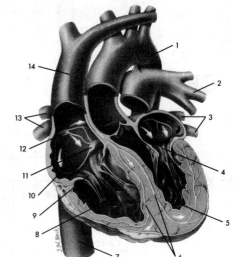

Fetal Circulation

1. Aortic arch
2. Ductus arteriosus
3. Pulmonary trunk
4. Abdominal aorta
5. Kidney
6. Common iliac artery
7. Internal iliac artery
8. Umbilical arteries
9. Umbilical cord
10. Fetal umbilicus
11. Umbilical vein
12. Hepatic portal vien
13. Liver
14. Ductus venosus
15. Inferior vena cava

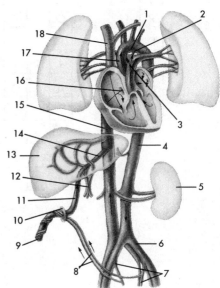

16. Foramen ovale
17. Ascending aorta
18. Superior vena cava

Hepatic Portal Circulation

1. Inferior vena cava
2. Stomach
3. Gastric vein
4. Spleen
5. Splenic vein
6. Gastroepiploic vein
7. Descending colon
8. Inferior mesenteric vein
9. Small intestine
10. Appendix
11. Ascending colon
12. Superior mesenteric vein
13. Pancreas
14. Duodenum
15. Hepatic portal vien
16. Liver
17. Hepatic veins

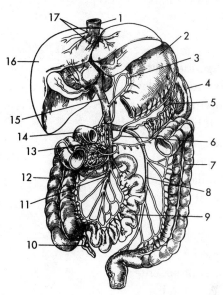

Principal Arteries of the Body

1. Occipital
2. Internal carotid
3. External carotid
4. Left common carotid
5. Left subclavian
6. Arch of aorta
7. Pulmonary
8. Left coronary
9. Aorta
10. Celiac
11. Splenic
12. Renal
13. Inferior mesenteric
14. Radial
15. Ulnar
16. Anterior tibial
17. Popliteal
18. Femoral
19. Deep femoral
20. External iliac
21. Internal iliac
22. Common iliac
23. Abdominal aorta
24. Superior mesenteric

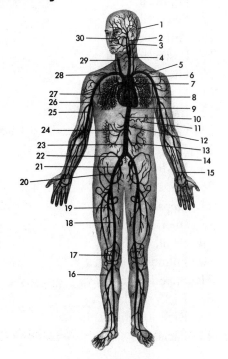

25. Brachial
26. Axillary
27. Right coronary
28. Brachiocephalic
29. Right common carotid
30. Facial

Principal Veins of the Body

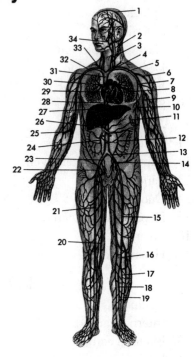

1. Superior sagittal sinus
2. External jugular
3. Internal jugular
4. Left brachiocephalic
5. Left subclavian
6. Cephalic
7. Axillary
8. Great cardiac
9. Basilic
10. Long thoracic
11. Splenic
12. Inferior mesenteric
13. Common iliac
14. Internal iliac
15. Femoral
16. Popliteal
17. Peroneal
18. Anterior tibial
19. Posterior tibial
20. Great saphenous
21. Femoral
22. External iliac
23. Common iliac
24. Superior mesenteric
25. Median cubital
26. Hepatic portal
27. Hepatic
28. Inferior vena cava
29. Small cardiac
30. Pulmonary (right)
31. Superior vena cava
32. Right subclavian
33. Right brachiocephalic
34. Facial

Normal ECG Deflections

1. Atrial depolarization
2. Ventricular depolarization
3. Ventricular repolarization

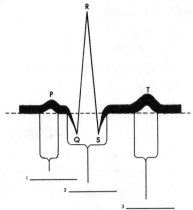

CHAPTER 12

THE LYMPHATIC SYSTEM AND IMMUNITY

Fill in the blanks

1. Lymph, p. 259
2. Interstitial fluid, p. 259
3. Lymphatic capillaries, p. 259
4. Right lymphatic duct and thoracic duct, p. 259
5. Cisterna chyli, p. 261
6. Lymph nodes, p. 261
7. Afferent, p. 262
8. Efferent, p. 262

Choose the correct response

9. B, p. 263
10. C, p. 263
11. C, p. 263
12. A, p. 262
13. C, p. 263
14. A, p. 262
15. A, p. 262

Matching

16. C, p. 263
17. A, p. 264
18. E, p. 264
19. B, p. 264
20. D, p. 265

Choose the correct term

21. C, p. 262
22. D, p. 262
23. E, p. 267

24. A, p. 265
25. H, p. 265
26. B, p. 265
27. I, p. 265
28. F, p. 266
29. J, p. 268
30. G, p. 265

Circle the term that does not belong

31. Allergy (all others refer to antibodies)
32. Complement (all others refer to antigens)
33. Antigen (all others refer to monoclonal antibodies)
34. Complement (all others refer to allergy)
35. Monoclonal (all others refer to complement)

Multiple choice

36. D, p. 268
37. D, p. 271
38. C, p. 272
39. B, p. 260
40. C, p. 271
41. C, p. 270
42. E, p. 271
43. E, p. 271
44. C, p. 271
45. E, p. 272
46. E, p. 273
47. D, p. 271
48. E, p. 269
49. A, p. 272
50. B, p. 273

Fill in the blanks

51. Stem cell, p. 270
52. Activated B cell, p. 271
53. Plasma cells, p. 271
54. Thymus gland, p. 273
55. Azidothymidine or AZT, p. 271
56. AIDS, p. 271
57. Vaccine, p. 271

Unscramble the words

58. Complement
59. Immunity
60. Clones
61. Interferon
62. Memory cells

APPLYING WHAT YOU KNOW

63. Interferon would possibly decrease the severity of the chickenpox virus.
64. AIDS
65. Baby Haskins had no means of producing T cells, thus making him susceptible to several diseases. Isolation was a means of controlling his exposure to these diseases.
66. WORD FIND

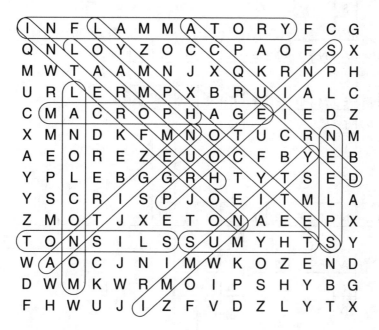

CROSSWORD

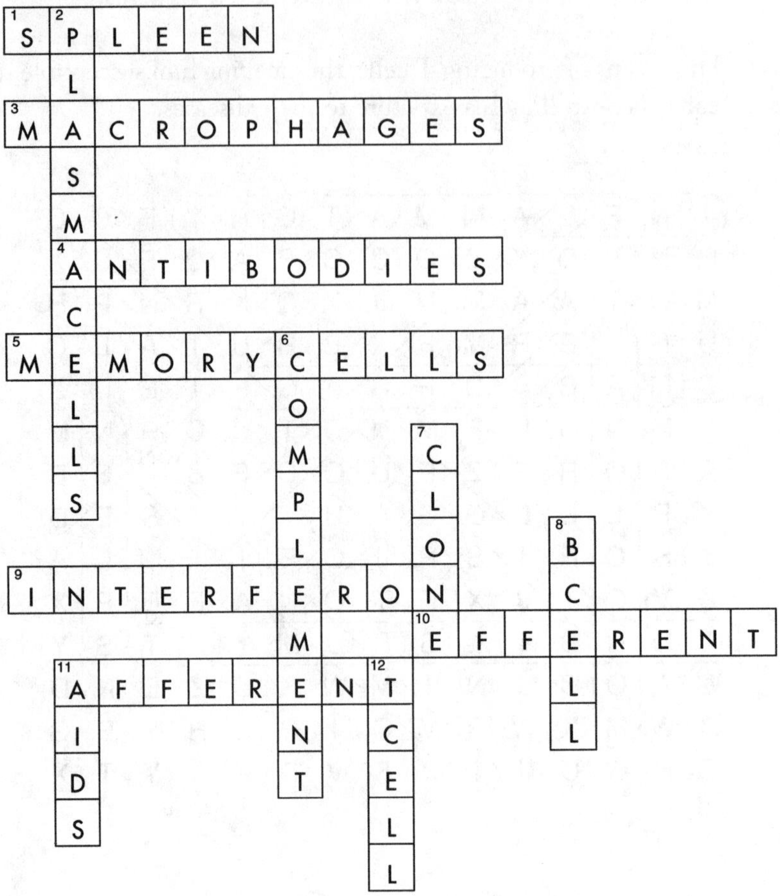
1 S 2 P L E E N
3 M A C R O P H A G E S
4 A N T I B O D I E S
5 M E M O R Y 6 C E L L S
7 C L O N
8 B C E L L
9 I N T E R F E R O N
10 E F F E R E N T
11 A F F E R E N 12 T
A I D S
C O M P L E M E N T
T C E L L
P L A S M A C E L L S

B Cell Development

1. Immature B cells
2. Activated B cells
3. Memory cells
4. Antibodies

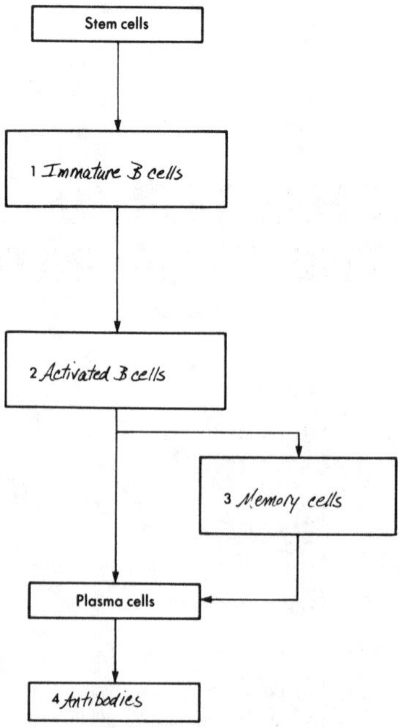
Stem cells
1 Immature B cells
2 Activated B cells
3 Memory cells
Plasma cells
4 Antibodies

Function of Sensitized T Cells

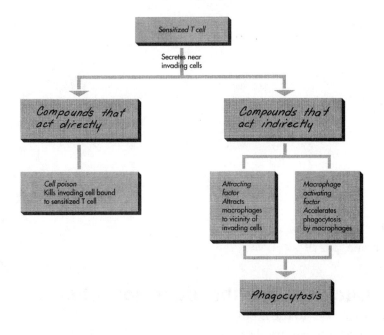

Principal Organs of Lymphatic System

1. Cervical lymph nodes
2. Thymas
3. Thoracic duct
4. Spleen
5. Cisterna chyli
6. Popliteal lymph nodes
7. Inguinal lymph nodes
8. Lymph vessels
9. Axillary lymph nodes
10. Submandibular nodes

CHAPTER 13
THE RESPIRATORY SYSTEM
Matching

1. J, p. 279
2. G, p. 279
3. A, p. 279
4. I, p. 282
5. B, p. 282
6. F, p. 282
7. C, p. 282
8. H, p. 282
9. D, p. 282
10. E, p. 279

Fill in the blanks

11. Air distributor, p. 279
12. Gas exchanger, p. 279
13. Filters, p. 279
14. Warms, p. 279
15. Humidifies, p. 279
16. Nose, p. 279
17. Pharynx, p. 279
18. Larynx, p. 279
19. Trachea, p. 279
20. Bronchi, p. 279
21. Lungs, p. 279
22. Alveoli, p. 279
23. Diffusion, p. 279
24. Respiratory membrane, p. 279
25. Surface, p. 279

Circle the one that does not belong

26. Oropharynx (the others refer to the nose)
27. Conchae (the others refer to paranasal sinuses)
28. Epiglottis (the others refer to the pharynx)
29. Uvula (the others refer to the adenoids)
30. Larynx (the others refer to the Eustachian tubes)
31. Tonsils (the others refer to the larynx)
32. Eustachian tube (the others refer to the tonsils)
33. Pharynx (the others refer to the larynx)

Choose the correct response

34. A, p. 283
35. B, p. 284
36. A, p. 283
37. A, p. 284
38. A, p. 283
39. B, p. 284
40. B, p. 284
41. C, p. 284

Fill in the blanks

42. Trachea, p. 284
43. Cartilage (C-rings), p. 286
44. Heimlich maneuver, pp. 286 and 287
45. Primary bronchi, p. 287
46. Alveolar sacs, p. 289
47. Apex, p. 290
48. Pleura p. 290
49. Pleurisy, p. 290
50. Pneumothorax, p. 290

True or false

51. Breathing, p. 290
52. Expiration, p. 293
53. Down, p. 293 (Review Chapter 2)
54. Internal respiration, p. 290
55. T, p. 295
56. 1 pint, p. 295
57. T, p. 295
58. Vital capacity, p. 295
59. T, p. 295

Multiple choice

60. E, p. 290
61. C, p. 295
62. C, p. 293
63. B, p. 292
64. D, p. 295
65. D, p. 295
66. D, p. 297

Matching

67. E, p. 297
68. B, p. 299
69. G, p. 299
70. A, p. 299
71. F, p. 299
72. D, p. 299
73. C, p. 299

Unscramble the words

74. Pleurisy
75. Bronchitis
76. Epistaxis
77. Adenoids
78. Inspiration

APPLYING WHAT YOU KNOW

79. During the day Mr. Gorski's cilia are paralyzed because of his heavy smoking. They use the time when Mr. Gorski is asleep to sweep accumulations of mucus and bacteria toward the pharynx. When he awakes, these collections are waiting to be eliminated.
80. Swelling of the tonsils or adenoids caused by infection may make it difficult or impossible for air to travel from the nose into the throat. The individual may be forced to breathe through the mouth.

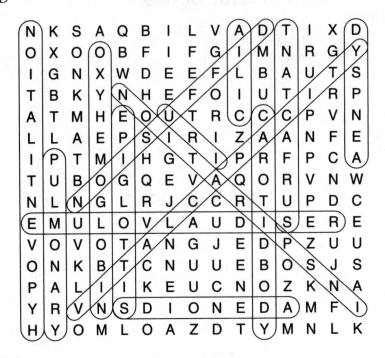

CROSSWORD

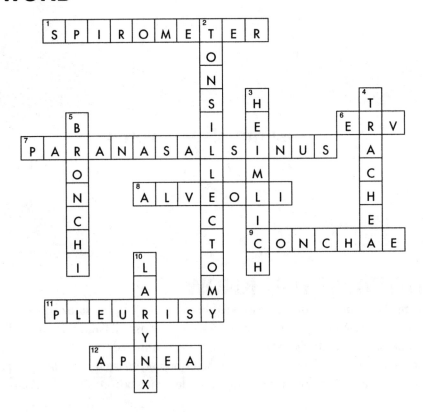

Sagittal View of Face and Neck

1. Sphenoid air sinus
2. Pharyngeal tonsil (adenoids)
3. Auditory tube
4. Soft palate
5. Uvula
6. Palatine tonsil
7. Lingual tonsil
8. Esophagus
9. Thyroid cartilage
10. Vocal cords
11. Epiglottis
12. Hyoid bone
13. Mandible
14. Tongue
15. Hard palate
16. Inferior concha
17. Middle concha
18. Nasal bone
19. Frontal air sinus
20. Superior concha

Respiratory Organs

1. Pharynx
2. Left primary bronchus
3. Bronchioles
4. Right primary bronchus
5. Trachea
6. Alveoli
7. Alveolar duct
8. Alveolar sac
9. Capillary

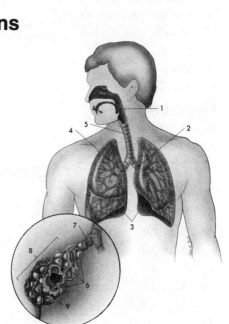

Pulmonary Ventilation Volumes

1. Total lung capacity
2. Inspiratory reserve volume
3. Tidal volume
4. Expiratory reserve volume
5. Residual volume

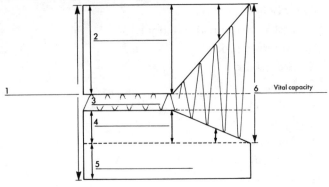

CHAPTER 14
THE DIGESTIVE SYSTEM

Fill in the blanks

1. Gastrointestinal tract or G.I. tract, p. 305
2. Mechanical, p. 305
3. Chemical, p. 305
4. Feces, p. 305
5. Digestion, absorption, and metabolism, p. 305
6. Parietal peritoneum, p. 305
7. Mouth, anus, p. 305
8. Lumen, p. 305
9. Mucosa, p. 305
10. Submucosa, p. 305
11. Peristalsis, p. 305
12. Serosa, p. 305
13. Mesentery, p. 305.

Choose the correct answer

14. A, p. 306
15. B, p. 306
16. B, p. 306
17. A, p. 306
18. A, p. 306
19. A, p. 306
20. A, p. 306
21. A, p. 306
22. B, p. 306
23. B, p. 306
24. B, p. 306
25. B, p. 306

Multiple choice

26. E, p. 308
27. C, p. 309
28. E, p. 309
29. D, p. 310

30. B, p. 310
31. C, p. 309
32. D, p. 309
33. D, p. 309
34. D, p. 311
35. A, p. 311
36. C, p. 311
37. A, p. 311
38. A, p. 309
39. B, p. 309
40. C, p. 309

Fill in the blanks

41. Pharynx, p. 312
42. Esophagus, p. 312
43. Stomach, p. 312
44. Cardiac sphincter, p. 312
45. Chyme, p. 312
46. Fundus, p. 313
47. Body, p. 313
48. Pylorus, p. 313
49. Pyloric sphincter, p. 313
50. Small intestine, p. 313

Matching

51. D, p. 312
52. J, p. 313
53. G, p. 312
54. A, p. 312
55. H, p. 314
56. B, p. 312
57. C, p. 312
58. E, p. 314
59. I, p. 314
60. F, p. 313

Multiple choice

61. C, p. 314
62. B, pp. 312 and 314 - 315
63. A, p. 317
64. A, p. 317
65. B, p. 314
66. E, p. 316
67. D, p. 314
68. D, p. 316; review Chapter 9 (hormones circulate in blood)
69. B, p. 317
70. C, p. 316

71. Vitamin K, p. 318
72. No villi are present in the large intestine, p. 319
73. Diarrhea, p. 319
74. Cecum, p. 320
75. Hepatic, p. 320
76. Sigmoid, p. 320
77. T, p. 320
78. T, p. 320
79. Parietal, p. 320
80. Mesentery, p. 321

Multiple choice

81. B, p. 322
82. D, p. 322
83. C, p. 322
84. C, p. 322
85. C, p. 322

86. Fill in the blank areas on the chart below.

CHEMICAL DIGESTION

DIGESTIVE JUICES AND ENZYMES	SUBSTANCE DIGESTED (OR HYDROLYZED)	RESULTING PRODUCT
SALIVA	1. Starch (polysaccharide)	
GASTRIC JUICE		2. Partially digested proteins
PANCREATIC JUICE		3. Peptides and amino acids
	4. Fats emulsified by bile	
	5. Starch	
INTESTINAL JUICE	6. Peptides	
7. Sucrase	8. Lactose	
		9. Glucose

APPLYING WHAT YOU KNOW

87. Ulcer
88. Pylorospasm
89. Basal metabolic rate or protein-bound iodine to determine thyroid function.

90. WORD FIND

CROSSWORD

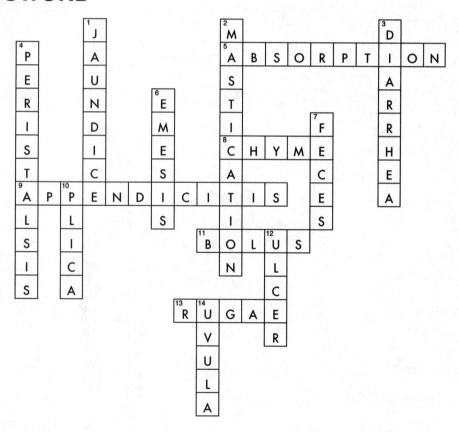

Digestive Organs

1. Parotid gland
2. Pharynx
3. Esophagus
4. Diaphragm
5. Spleen
6. Splenic flexure
7. Stomach
8. Transverse colon
9. Descending colon
10. Ileum
11. Sigmoid colon
12. Rectum
13. Anal canal
14. Vermiform appendix
15. Cecum
16. Ileocecal valve
17. Ascending colon
18. Pancreas
19. Duodenum
20. Common bile duct
21. Hepatic bile duct
22. Hepatic flexure
23. Cystic duct
24. Gallbladder
25. Liver
26. Trachea
27. Larynx
28. Submandibular gland
29. Sublingual gland
30. Tongue

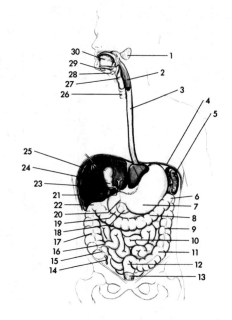

Tooth

1. Enamel
2. Pulp
3. Pulp cavity
4. Gingiva (gum)
5. Dentin
6. Periodontal membrane
7. Cementum
8. Root canal
9. Vein and artery
10. Nerve
11. Root
12. Neck
13. Crown

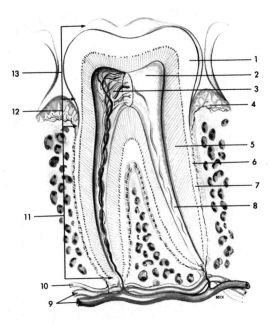

The Salivary Glands

1. Parotid duct
2. Parotid gland
3. Submandibular duct
4. Submandibular gland
5. Sublingual gland

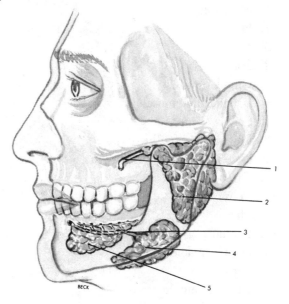

Stomach

1. Fundus
2. Body
3. Greater curvature
4. Rugae
5. Pylorus
6. Pyloric sphincter
7. Duodenum
8. Lesser curvature
9. Oblique muscle layer
10. Circular muscle layer
11. Longitudinal muscle
12. Cardiac sphincter
13. Esophagus

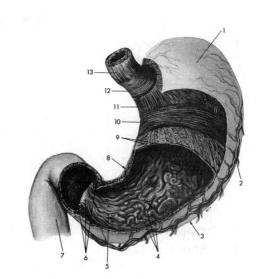

Gallbladder and Bile Ducts

1. Right and left hepatic ducts
2. Common hepatic duct
3. Common bile duct
4. Accessory duct
5. Pancreas
6. Pancreatic duct
7. Duodenum
8. Major duodenal papilla
9. Sphincter muscles
10. Minor duodenal papilla
11. Gallbladder
12. Cystic duct

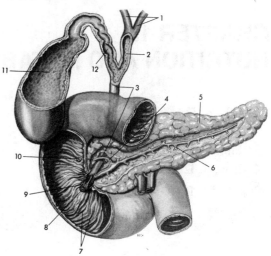

The Small Intestine

1. Mesentery
2. Plica
3. Mucosa
4. Submucosa
5. Circular muscle
6. Longitudinal muscle
7. Serosa
8. Epithelium of villus
9. Lacteal
10. Artery
11. Vein
12. Plica
13. Submucosa
14. Lymph nodules
15. Serosa
16. Circular muscle
17. Longitudinal muscle

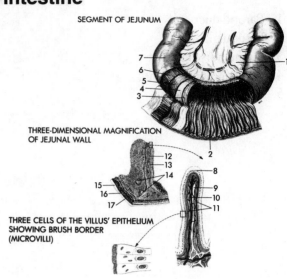

SEGMENT OF JEJUNUM

THREE-DIMENSIONAL MAGNIFICATION OF JEJUNAL WALL

THREE CELLS OF THE VILLUS' EPITHELIUM SHOWING BRUSH BORDER (MICROVILLI)

The Large Intestine

1. Splenic flexure
2. Descending colon
3. Sigmoid colon
4. Rectum
5. Vermiform appendix
6. Ileum
7. Cecum
8. Ileocecal valve
9. Ilium
10. Ascending colon
11. Hepatic flexure
12. Inferior vena cava
13. Aorta
14. Transverse colon

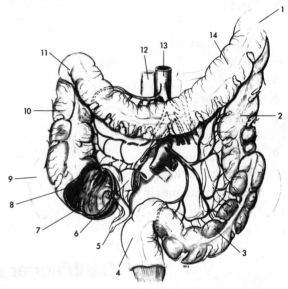

CHAPTER 15
NUTRITION AND METABOLISM

Fills in the blanks

1. Bile, p. 329
2. Prothrombin, p. 329
3. Fibrinogen, p. 329
4. Iron, p. 329
5. Hepatic portal vein, p. 329

Matching

6. B, p. 332
7. A, p. 329
8. C, p. 332
9. D, p. 333
10. E, p. 333
11. A, p. 329
12. E, p. 333
13. A, p. 329

Circle the term that does not belong

14. Bile (all others refer to carbohydrate metabolism)
15. Amino acids (all others refer to fat metabolism)
16. M (all other refer to vitamins)
17. Iron (all others refer to protein metabolism)
18. Insulin (all others tend to increase blood glucose)
19. Folic acid (all others are minerals)
20. Ascorbic acid (all others refer to the B-complex vitamins)

Circle the correct choice

21. C, p. 334
22. A, p. 334
23. C, p. 336
24. B, p. 336
25. B, p. 336
26. A, p. 336
27. C, p. 336
28. D, p. 336
29. A, p. 336

Unscramble the words

30. Liver
31. Catabolism
32. Amino
34. Pyruvic
35. Evaporation

APPLYING WHAT YOU KNOW

35. Weight loss; Anorexia nervosa
36. Iron

    Meat, eggs, vegetables and legumes
37. He was carbohydrate loading or glycogen loading, which allow the muscles to sustain aerobic exercise for up to 50% longer than usual.

CROSSWORD

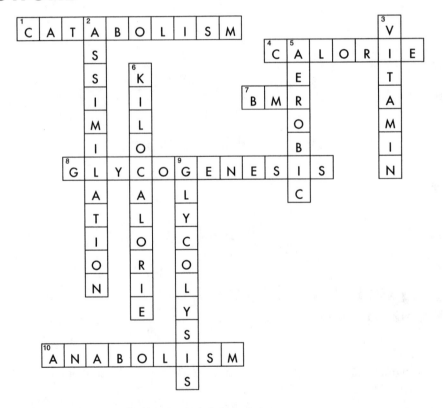

CHAPTER 16
URINARY SYSTEM

Multiple choice

1. E, p. 341
2. C, p. 343
3. E, p. 343
4. C, p. 345
5. E, p. 347
6. C, p. 347
7. B. p. 347
8. E, p. 341, 345
9. C, p. 348
10. C, p. 348
11. B, p. 348 (review Chapter 9)
12. D, p. 348

Choose the correct term

13. G, p. 341
14. I, p. 349
15. H, p. 341
16. B, p. 341
17. K, p. 343
18. F, p. 341
19. J, p. 343
20. D, p. 341
21. L., p. 343
22. C, p. 341
23. M, p. 351
24. A, p. 341

Choose the correct term

25. B, p. 352
26. C, p. 353
27. A, p. 349
28. B, p. 352
29. C, p. 353
30. C, p. 353
31. A, p. 349
32. C, p. 352 Fig. 16-8
33. B, p. 352
34. A, p. 349
35. B, p. 352

Fill in the blanks

36. Renal colic, p. 352
37. Mucous membrane, p. 349
38. Renal calculi, p. 354

39. Ultrasound, p. 354
40. Catheterization, p. 353
41. Renal pelvis, p. 349
42. Semen, p. 353
43. Urinary meatus, p. 353

Fill in the blanks

44. Micturition, p. 353
45. Urination, p. 353
46. Voiding, p. 353
47. Internal urethral, p. 353
48. Exit, p. 353
49. Urethra, p. 353
50. Voluntary, p. 353
51. Emptying reflex, p. 353
52. Urethra, p. 353
53. Retention, p. 353
54. Suppression, p. 354
55. Automatic bladder, p. 354

APPLYING WHAT YOU KNOW

56. Polyuria
57. Residual urine is often the cause of repeated cystitis
58. A high percentage of catheterized patients develop cystitis, often due to poor aseptic technique when inserting the catheter.
59. WORD FIND

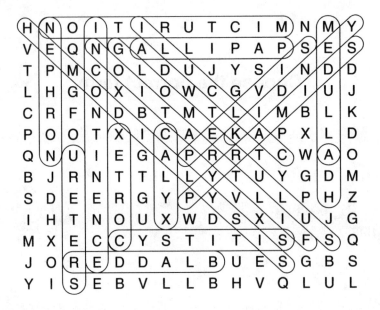

CROSSWORD

Across:
3. CYSTITIS
7. ANURIA
8. CATHETERIZATION
11. GLOMERULUS

Down:
1. MICTURITION
2. LITHOTRIPTOR
3. CALYX
4. INCONTINENCE
5. TRIGONON
6. GLYCOSURIA
9. POLYURIA
10. OLIGURIA

Urinary System

1. Spleen
2. Renal artery
3. Renal vein
4. Kidney (left)
5. Inferior vena cava
6. Abdominal aorta
7. Common iliac vein
8. Common iliac artery
9. Urinary bladder
10. Ureter
11. Kidney (right)
12. Tenth rib
13. Adrenal glands
14. Liver

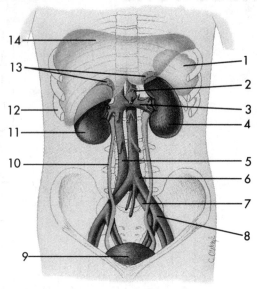

Kidney

1. Renal artery and vein
2. Pelvis
3. Ureter
4. Cortex
5. Pyramid
6. Medulla
7. Calyx
8. Papilla

Nephron

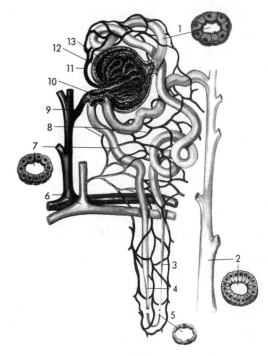

1. Proximal convoluted tubule
2. Collecting tubule
3. Descending limb of Henle's loop
4. Ascending limb of Henle's loop
5. Segment of Henle's loop
6. Artery and vein
7. Distal convoluted tubule
8. Peritubular capillaries
9. Afferent arteriole
10. Juxtaglomerular apparatus
11. Efferent arteriole
12. Glomerulus
13. Bowman's capsule

CHAPTER 17
FLUID AND ELECTROLYTE BALANCE

Circle the correct response

1. Inside, p. 359
2. Extracellular, p. 359
3. Extracellular, p. 359
4. Lower, p. 359
5. More, p. 359
6. Decline, p. 359
7. Less, p. 359
8. Decreases, p. 359
9. 55%, p. 359
10. Fluid balance, p. 359

Multiple choice

11. A, p. 364
12. D, p. 364
13. A, p. 364
14. A, p. 364
15. C, p. 361
16. E, p. 361
17. D, p. 362
18. D, p. 365
19. C, p. 367
20. B, p. 365
21. D, p. 363

22. E, p. 366
23. B, p. 367
24. B, p. 367
25. B, p. 367
26. E, p. 364

True or false

27. Catabolism, p. 359
28. T, p. 359
29. T, p. 362
30. Nonelectrolyte, p. 362
31. T, p. 364
32. Hypervolemia, p. 365
33. Tubular function, p. 363
34. 2400 ml, p. 361
35. T, p. 367
36. 100 mEq, p. 365

Fill in the blanks

37. Dehydration, p. 367
38. Decreases, p. 367
39. Decrease, p. 367
40. Overhydration, p. 367
41. Intravenous fluids, p. 367
42. Heart, p. 367

APPLYING WHAT YOU KNOW

43. Ms. Titus could not accurately measure water intake created by foods or catabolism, nor could she measure output created by lungs, skin, or the intestines.
44. A careful record of fluid intake and output should be maintained and the patient should be monitored for signs and symptoms of electrolyte and water imbalance.

45. WORD FIND

```
S T V H O M E O S T A S I S H
I E D E M A S Q A L P U E G L
M M L U F L U I D O Y O I S L
W S B E A E C O L D P N I E R
L I T A C V S H Q O C E H B C
L L X J L T E A W Y B V Q Q O
I O P E A R U C T C A A R I N
N B H R X S N O I I P R T C N
S A O X M G J C L N J T B A H
I N X X D V U D E Y K N Y O C
E A A J Q D I U R E T I C S X
I F C J A M P V E N B E Y F W
V F K T Q X R M D D S I V O T
E T T Z N T R P X I M L J F I
S W Y A P V Q N S K T K W B P
```

CROSSWORD

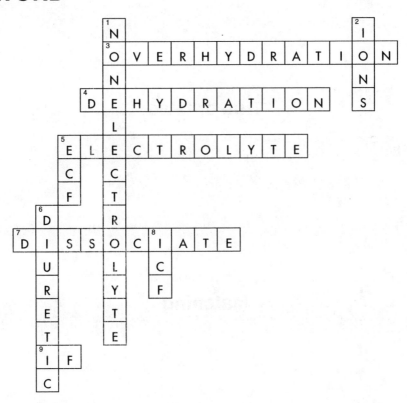

CHAPTER 18
ACID-BASE BALANCE

Choose the correct term

1. B, p. 371
2. A, p. 371
3. A, p. 371
4. B, p. 371
5. B, p. 371
6. B, p. 371
7. B, p. 373
8. A, p. 372
9. B, p. 372
10. B, p. 372

Multiple choice

11. E, p. 371
12. E, p. 371
13. A, p. 371
14. E, p. 372
15. C, p. 375
16. D, p. 375
17. C, p. 375
18. B, p. 375
19. D, p. 376
20. E, p. 376
21. E, p. 376

True or false

22. Buffer instead of heart, p. 371
23. Buffer pairs, p. 372
24. T, p.372
25. T, p. 374
26. Alkalosis, p. 375
27. Kidneys, p. 375
28. T, p. 375
29. Kidneys, p. 377
30. Lungs, p. 377

Matching

31. E, p. 378
32. G, p. 378
33. F, p. 378
34. A, p. 378
35. I, p. 379
36. B, p. 378
37. H, p. 378
38. C, p. 378
39. D, p. 378
40. J, p. 378

APPLYING WHAT YOU KNOW

41. Normal saline contains chloride ions that replace bicarbonate ions and thus relieve the bicarbonate excess that occurs during severe vomiting.

42. Most citrus fruits, although acid tasting, are fully oxidized during metabolism and have little effect on acid-base balance. Cranberry juice is one of the few exceptions.

43. Milk of magnesia. It is base. Milk is slightly acidic.

44. WORD FIND

CROSSWORD

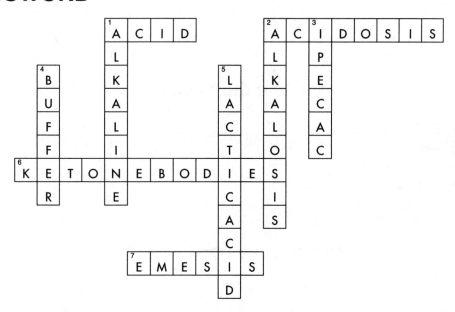

CHAPTER 19
THE REPRODUCTIVE SYSTEMS
Matching

Group A
  1. D, p. 384
  2. C, p. 383
  3. E, p. 383
  4. B, p. 384
  5. A, p. 383

Group B
  6. C, p. 384
  7. A, p. 384
  8. D, p. 383
  9. B, p. 384
10. E, p. 384

Multiple choice

11. B, p. 385
12. C, p. 385
13. A, p.385
14. D, p.386
15. E, p. 385
16. D, p. 388
17. C, p. 388, 389
18. C, p. 386
19. A, p. 388
20. B, p. 386

Fill in the blanks

21. Testes, p. 384
22. Spermatozoa or sperm, p. 386
23. Ovum, p. 383
24. Testosterone, p. 388
25. Interstitial cells, p. 388
26. Masculinizing, p. 389
27. Anabolic, p. 389

Choose the correct term

28. B, p. 389
29. H, p. 391
30. G, p. 389
31. A, p. 389
32. F, p. 389
33. C, p. 389
34. I, p. 389

35. E, p. 389
36. D, p. 390
37. J, p. 390

Matching

38. D, p. 391
39. C, p. 391
40. B, p. 391
41. A, p. 391
42. E, p. 391

Choose the correct structure

43. A, p. 397
44. B, p. 396
45. A, p. 397
46. B, p. 394
47. A, p. 397
48. A, p. 397
49. A, p. 397
50. B, p. 392

Fill in the blanks

51. Gonads, p. 392
52. Oogenesis, p. 393
53. Meiosis, p. 393
54. One half or 23, p. 393
55. Fertilization, p. 393
56. 46, p. 393
57. Estrogen, p. 393
58. Progesterone, p. 393
59. Secondary sexual characteristics, p. 394
60. Menstrual cycle, p. 394
61. Puberty, p. 394

Choose the correct structure

62. A, p. 394
63. B, p. 395
64. C, p. 396
65. B, p. 395
66. A and C, p. 394
67. B, p. 394
68. A, p. 394
69. A, p. 394
70. C, p. 396
71. B, p. 400

Matching

Group A
72. D, p. 396

73. E. p. 396
74. B, p. 396
75. C, p. 396
76. A, p. 396

Group B
77. E, p. 396
78. A, p. 397
79. D, p. 397
80. B, p. 397
81. C, p. 397

True or false

82. Menarche, p. 397
83. One, p. 400
84. 14, p. 400
85. Menstrual period, p. 400
86. T, p. 400
87. Anterior, p. 400

Choose the correct hormone

88. B, p. 400
89. A, p. 400
90. B, p. 400
91. B, p. 400
92. A, p. 401

APPLY WHAT YOU KNOW

93. Yes. The testes are not only essential organs of reproduction, but are also responsible for the "masculinizing" hormone. Without this hormone, Mr. Belinki will have no desire to reproduce.
94. Sterile—The sperm count may be too low to reproduce but the remaining testicle will produce enough masculinizing hormone to prevent impotency.
95. The uterine tubes are not attached to the ovaries and infections can exit at this area and enter the abdominal cavity.
96. Yes. Yes. Without the hormones from the ovaries to initiate the menstrual cycle, Mrs. Harlan will no longer have a menstrual cycle and can be considered to be in menopause (cessation of menstrual cycle).
97. No. Ms. Comstock still will have her ovaries, which are the source of her hormones, She will not experience menopause because of this procedure.

CROSSWORD

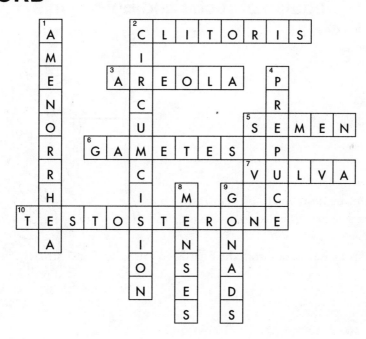

Male Reproductive Organs

1. Ductus deferens
2. Bladder
3. Prostate gland
4. Urethra
5. Penis
6. Testis
7. Epididymis
8. Bulbourethral gland
9. Anus
10. Seminal vesicle
11. Rectum

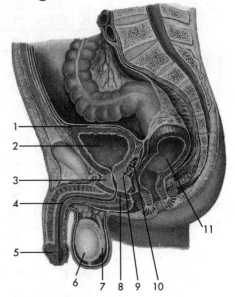

Tubules of Testis and Epididymis

1. Ductus (vas) deferens
2. Body of epididymis
3. Septum
4. Lobule
5. Tunica albuginea
6. Seminiferous tubules

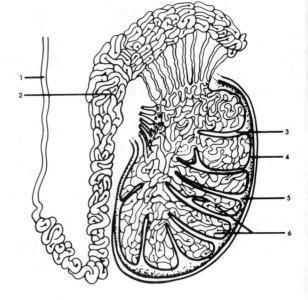

Vulva

1. Mons pubis
2. Clitoris
3. Orifice of urethra
4. Labia majora
5. Opening of greater vestibular gland
6. Anus
7. Vestibule
8. Orifice of vagina
9. Labia minora
10. Prepuce

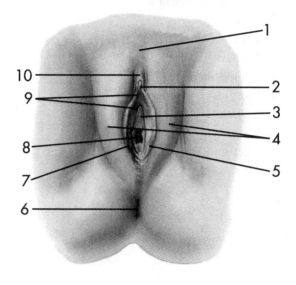

Breast

1. Clavicle
2. Pectoralis major muscle
3. Intercostal muscles
4. Nipple
5. Lactiferous duct
6. Alveoli
7. Suspensory ligaments of Cooper

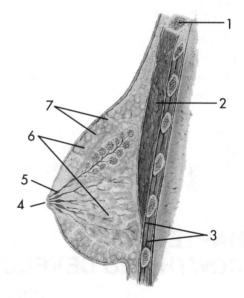

Female Pelvis

1. Fallopian tube (uterine)
2. Ovary
3. Uterus
4. Urinary bladder
5. Symphysis pubis
6. Urethra
7. Vagina
8. Cervix
9. Rectum

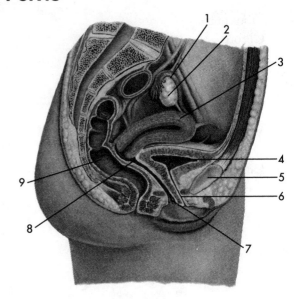

Uterus and Adjacent Structures

1. Fundus
2. Endometrium
3. Myometrium
4. Cervical canal
5. Vagina
6. Cervix
7. Body
8. Uterine (fallopian) tube
9. Ovary
10. Fimbriae

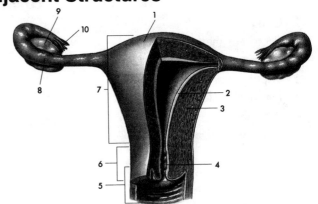

CHAPTER 20
GROWTH AND DEVELOPMENT

Fill in the blanks

1. Conception, p. 407
2. Birth, p. 407
3. Embryology, p. 407
4. Oviduct, p. 407
5. Zygote, p. 407
6. Morula, p. 407
7. Blastocyst, p. 407
8. Amniotic cavity, p. 409
9. Chorion, p. 411
10. Placenta, p. 411

Choose the correct term

11. G, p. 410
12. F, p. 412
13. C, p. 416
14. B, p. 412
15. A, p. 410
16. H, p. 414
17. E, p. 414
18. D, p. 414
19. I, p. 412
20. J, p. 416

Multiple choice

21. E, p. 416, 417
22. E, p. 417
23. E, p. 418
24. A, p. 418
25. B, p. 418
26. C, p. 418
27. B, p. 418
28. D, p. 418
29. E, p. 418
30. D, p. 419
31. A, p. 419
32. C, p. 419
33. C, p. 420
34. C, p. 419
35. E, p. 419

Matching

36. F, p. 416
37. A, p. 417
38. C, p. 419
39. H, p. 419
40. D, p. 419
41. B, p. 417
42. E, p. 419
43. G, p. 417
44. I, p. 420

Fill in the blanks

45. Lipping, p. 420
46. Osteoarthritis, p. 420
47. Nephron, p. 420
48. Barrel chest, p. 420
49. Atherosclerosis, p. 422
50. Arteriosclerosis, p. 422
51. Hypertension, p. 422
52. Presbyopia, p. 422

53. Cataract, p. 422
54. Glaucoma, p. 422

Unscramble the words

55. Infancy
56. Postnatal
57. Organogenesis
58. Zygote
59. Childhood
60. Fertilization

APPLYING WHAT YOU KNOW

61. Normal
62. Only about 40% of the taste buds present at age 30 remain at age 75
63. A significant loss of hair cells in the Organ of Corti cause a serious decline in ability to hear certain frequencies
64. WORD FIND

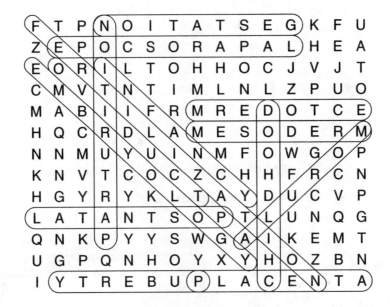

CROSSWORD

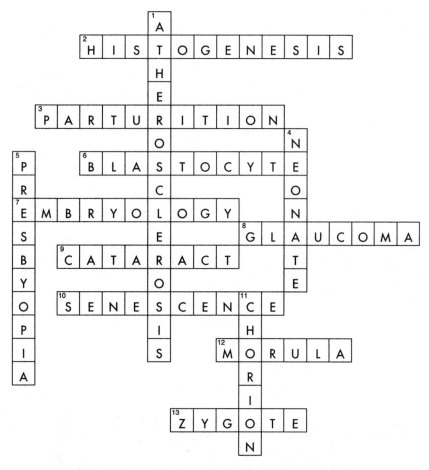

Fertilization and Implantation

1. Developing follicles
2. Corpus luteum
3. Ovary
4. Ovulation
5. Spermatozoa (fertilization)
6. First mitosis
7. Uterine (Fallopian) tube
8. Divided zygote
9. Morula
10. Uterus
11. Blastocyst
12. Implantation

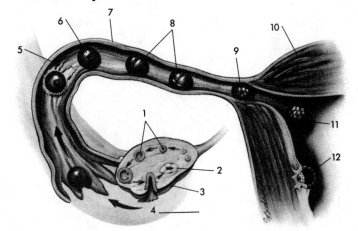